SOIL REMEDIATION FOR THE PETROLEUM EXTRACTION INDUSTRY

SECOND EDITION

SOIL REMEDIATION FOR THE PETROLEUM EXTRACTION INDUSTRY

SECOND EDITION

Lloyd E. Deuel, Jr., Ph.D.

George H. Holliday, Ph.D., P. E. DEE

PennWell

PennWell
1421 South Sheridan/P.O. Box 1260
Tulsa, Oklahoma 74101

ISBN 0-87814-740-3

Printed in the United States of America

1 2 3 4 5 99 98

CONTENTS

Preface

Although soil science has been applied to the production of agricultural crops for many years, few people have applied soil science to remediating contaminated soil and non-hazardous oilfield wastes (NOW) found in oil and gas extraction operations. This book was written to present to the readers means of applying soil science technology to oil and gas field remediation and pit closure problems. The book provides specific recommendations for reserve pit management so as to avoid the need to treat the pit wastes. Additionally, it contains necessary formulae, criteria and soil amendments information so the reader can apply the demonstrated technology to field problems. Required soil analytical methods are presented in detail in the Appendices. These methods are written in sufficient detail so a laboratory inexperienced in soil analysis can obtain useable and repeatable analytical results.

The first edition of this book has been used for three years as the text for a two-day short course in Soil Remediation for training oil and gas company personnel and state/federal personnel. The authors gratefully acknowledge the corrections and critiques offered by the many students attending the short courses. This second edition attempts to incorporate the many suggestions offered by the students and readers. We owe them a great debt of gratitude for their helpful and constructive criticisms.

Further, we express our thanks to Shawn Hokanson, who is responsible for upgrading the figures and color reproduction found in the second edition. His computer expertise has improved and enhanced the visual aspects of the book and has been invaluable to the authors.

George Holliday
Bellaire, TX

Lloyd E. Deuel, Jr.
College Station, TX

FIGURES AND CAPTIONS

Color Plates

LIST OF TABLES

1

GLOSSARY OF SOIL SCIENCE TERMS

The following listing compiles terms and definitions used for describing soils and solid wastes.

TERMS

Aeration, Soil. The process by which air in the soil is replaced by air from the atmosphere. In a well-aerated soil, the soil air is very similar in composition to the atmosphere above the soil. Poorly aerated soils usually contain a much higher percentage of carbon dioxide and a correspondingly lower percentage of oxygen than the atmosphere above the soil.

Air Porosity. The bulk volume of air in soil at any given time or under a given condition, such as a specified moisture tension (negative pressure).

Available Water. The portion of water in a soil readily absorbed by plant roots. Usually described as water held in the soil against a pressure between 1/3 and 15 atmospheres (220 psi).

Bulk Density, Soil. The mass of dry soil per unit of bulk volume. The bulk volume is determined before drying at 105°C to constant weight.

Cation Exchange Capacity (CEC). The total of the exchangeable cations a soil can adsorb, expressed in milliequivalents per 100 grams or per gram of soil. The major cations are calcium, magnesium, potassium and sodium.

Clay Mineral. (1) Naturally occurring inorganic crystalline material found in soils and other earthy deposits, usually of clay size (less than 0.002 mm in diameter). (2) Material as described under (1), but not limited by particle size.

Clay Particle. (1) A soil separate consisting of particles less than 0.002 mm in equivalent diameter. (2) Clay as a unit describes a textural class.

Crust. A surface layer of soil, ranging in thickness from a few millimeters to perhaps as much as an inch, that is much more compact, hard, and brittle when dry than the material immediately beneath it.

Deflocculate. (1) To separate compound particles into individual components by chemical and/or physical means. (2) To cause the particles of the disperse phase of a colloidal system to become suspended in the dispersion medium.

Disperse. (1) To break up compound particles, such as aggregates, into the individual component particles. (2) To distribute or suspend fine particles, such as clay, in or throughout a dispersion medium, such as water.

Dry-Weight Percentage. The ratio of the weight of any constituent of a soil to the oven-dry weight of the soil (constant weight at 105°C).

Electrical Conductivity (EC). Conductivity measured directly in reciprocal units of resistance and reported in millimhos per centimeter (mmhos/cm). EC is an indirect measure of total dissolved solids (TDS). EC can be determined from saturated paste, 1:1 by weight or 1:1 by volume extracts.

Exchange Capacity. The total charge of the adsorption complex active in the adsorption of ions.

Exchangeable Cation Percentage. The extent to which the adsorption complex of a soil is occupied by a particular cation.

Exchangeable Sodium Percentage (ESP). The percentage of the cation exchange capacity of a soil occupied by sodium.

Fertility, Soil. The status of a soil with respect to the amount and availability of elements necessary for plant growth.

Field Capacity. Saturated soil water content available at a tension of 1/3 atmosphere.

Fine Texture. Soil consisting of or containing large quantities of fine fractions, particularly of silt and clay.

Gravitational Water. Water which moves into, through or out of the soil under the influence of gravity.

Groundwater. That portion of the total precipitation which at any particular time is either passing through or standing in the soil and the underlying strata; which is free to move under the influence of gravity.

Halophyte. A plant that grows in salty soils; a plant thriving as opposed to tolerating a saline environment.

Heavy Soil. A soil having a high content of fine separates, such as clay. Synonymous with fine textured soil.

Humin. The fraction of the organic matter in a soil not dissolved upon extraction of the soil with dilute alkali.

Humus. The more or less stable fraction of organic matter in the soil remaining after the major portion of added plant and animal residues have decomposed.

Infiltration. The downward entry of water into the soil.

Infiltration Rate. A soil characteristic determining or describing the maximum rate at which water can enter the soil under specified conditions, including the presence of or an excess of water.

Kaolinite. (1) An aluminosilicate mineral of the 1:1 crystal lattice group; that consists of one silicon tetrahedral layer and one aluminum oxide-hydroxide octahedral layer. (2) The 1:1 group of the family of aluminosilicates.

Labile. Refers to a substance in soil readily transformed or presently available to plants.

Land Classification. The arrangement of land units into various categories based upon the properties of the land or its suitability for some particular purpose, e.g., farmland.

Land Spreading. A process in which contaminated soils or wastes are spread over a treatment area and tilled with native soil.

Land Treatment. A process in which contaminated soils or wastes are spread over a treatment area and tilled with native soil. Nutrients, water, and air are added to enhance biodegradation.

Landscape. All the natural features such as fields, hills, forests, water, etc., which distinguish one part of the earth's surface from another part.

Lime, Agricultural. A soil amendment consisting principally of calcium carbonate, but including magnesium carbonate; used to furnish calcium and magnesium and neutralize soil acidity.

Macronutrient. A chemical element necessary in large amounts for the normal growth of plants and usually applied artificially in fertilizer or liming material.

Marsh. Periodically wet or continually flooded areas where the surface is not deeply submerged. Subclasses include freshwater and saltwater marshes.

Micronutrient. A chemical element necessary in extremely small amounts for the normal growth of plants. Examples are: Boron (B), Chloride (Cl), Copper (Cu), Iron (Fe), Manganese (Mn), and Zinc (Zn).

Moisture Tension (or Pressure). The negative pressure in the soil water. It is equal to the pressure applied to soil water to bring it to hydraulic equilibrium through a permeable wall or membrane, with a pool of water of the same composition. The pressures used and the corresponding percentages most commonly determined are: (1) Fifteen-atmosphere- percentage - The percentage of water contained in a saturated soil subjected to an applied pressure of 15 atm until it is in equilibrium. (2) One-third-atmosphere percentage - The percentage of water contained in a saturated soil which has been subjected to an applied pressure of 1/3 atm until it is in equilibrium.

Montmorillonite. An aluminosilicate clay mineral having a 2:1 expanding crystal lattice, e.g., consisting of two silicon tetrahedral layers enclosing an aluminum octahedral layer. Considerable expansion may be caused by water moving between silica layers of contiguous units (interlayer water absorption).

Muck Soil. (1) A soil containing between 20 and 50% organic matter. (2) An organic soil in which the organic matter is well decomposed.

Nitrogen (N). Chemical element necessary for plant growth.

Oil. Petroleum hydrocarbons in crude or refined products.

Oven Dry Soil. Soil dried at 105°C until it reaches constant weight.

Particle Density. The mass per unit volume of the soil particles.

Permeability, Soil. The ease with which gases, liquids or plant roots penetrate or pass through a bulk mass or a layer of soil.

Physical Properties of Soils. Those characteristics, processes, or reactions of a soil which are caused by physical forces and which can be described by equation or expressed in physical terms.

Phosphorous (P). Chemical element necessary for plant growth.

Porosity. The percentage of the total volume not occupied by soil particles.

Potassium (K). Chemical element necessary for plant growth.

Saline Soil. A nonsodic soil containing sufficient soluble salts to impair its productivity. The electrical conductivity of the saturation extract is greater than 4 millimhos per centimeter (at 25°C) and the pH is usually less than 8.3 units.

Saline Sodic. (1) A soil containing sufficient exchangeable sodium to interfere with the growth of most crop plants and also containing appreciable quantities of soluble salts. (2) A soil having an exchangeable sodium percentage (ESP) greater than 15% and the electrical conductivity (EC) of the saturation extract is greater than 4 millimhos per centimeter.

Sand. A soil particle between 0.05 and 2.0 mm in diameter.

Saturation. (1) A condition in which all the voids between soil particles are filled with a liquid. (2) A condition in which the most concentrated solution possible is formed under a given set of physical conditions.

Saturated Paste (SP) Extract. A liquid sample representative of a saturated soil.

Silt. A soil separate consisting of particles between 0.05 and 0.002 mm in equivalent diameter.

Sodic Soil. A soil that contains an exchangeable sodium percentage (ESP) of 15% or more.

Sodium Adsorption Ratio (SAR). An empirical mathematical expression developed as an index to the sodium hazard in soils:
$SAR = Na / [(Ca + Mg)/2]^{1/2}$
High Na levels (SAR>12) in soil solution cause Ca and Mg deficiencies in plants.

Soil. (1) The unconsolidated mineral material on the immediate surface of the earth serving as natural medium for the growth of land plants. (2) The unconsolidated mineral matter on the surface of the earth having been subjected to and influenced by genetic and environmental factors of: parent material, climate, macro- and microorganisms and topography. These factors act over a period of time to produce a product-soil differing from the matter from which it is derived in many physical, chemical, biological and morphological properties and characteristics.

Solum. The upper and most weathered part of the soil profile.

Special Management. Procedures used on pit or spill wastes requiring more complex closure procedures than simply mixing waste fluids with levee materials to complete closure.

Surface Soil. The uppermost part of the soil, ordinarily moved in tillage operations.

Swamp. An area saturated with water throughout much of the year; usually the surface of the soil is not deeply submerged.

Tidal Flats. Areas that are nearly flat, barren, and periodically covered by tidal waters.

Total Dissolved Solids (TDS). The total concentration of dissolved cations (mg/l) in the soil moisture.

Total Petroleum Hydrocarbons (TPH). The total hydrocarbons present in soil as determined by a particular analytical method.

Wasteland. Land not suitable for, or capable of producing materials or services of value.

Wilting Point. Water held in the soil against a pressure of 15 atmospheres.

SOIL AND PHYSICAL RELATIONSHIPS

The following conversion factors and equations provide a ready means of calculating quantities of soil amendments required for proper remediation.

Conversions

5.61 ft^3 = 1 bbl
Carbon (mg/kg) = % TPH x .78 x 10,000 = C
TDS (mg/liter) for saturated paste extract = EC (mmhos/cm) x 613
For oil bioremediation:
C/N ratio = 150/1
P/N and K/N ratio = 1/4
Moisture content = 80% of field capacity ≅ 1″ of water per week
Ammonium nitrate = 33% N
Super phosphate = 21% P_2O_5, or 9.2% P
Muriate of Potash = 61% K_2O, or 51% K
12-12-12 fertilizer means 12% Nitrogen (N), 12% P_2O_5 [5.3% (0.44 x 12%) actual Phosphorous (P)], and 12% K_2O [10% (0.83 x 12%) actual Potassium (K)]
21,780 ft^3 = 1 acre-6 inches deep
1 acre-6 inch = 2,000,000 lb soil
1 ft^3 soil = 91.8 lb

Equations

SAR = Soluble Na(meq/l)/[{Soluble Ca(meq/l) + Soluble Mg(meq/l)}/2]$^{1/2}$

ESP % = [Exchangeable Na(meq/100g)/CEC(meq/100g)] * 100

Relationship of ESP to SAR

ESP = [100 (-0.0216 + 0.01475) SAR] / [1+(-0.0216 + 0.01475) SAR]

EC Calculation

Desired EC = [(contaminated vol.) × (measured EC)+(x background vol.) × background EC)]/(contaminated vol.+ x)

Where x (ft^3) = background soil volume needed for dilution

Solving for x:

$$x = \frac{[(\text{measured EC})-(\text{desired EC})](\text{contaminated volume})}{(\text{desired EC - background EC})}$$

Total Volume (ft^3) = Contaminated volume + Dilution volume

This equation works for any contamination you have, such as TPH, SAR, ESP, etc.:

SPEC = [(1:1 by weight EC) × 100]/[SP moisture percent]

Where SPEC = Saturated paste extract EC

Gypsum and Sulfur Requirement Determined from CEC and ESP

[(% ESP-12%)/100] × (CEC)	=	charge requirement (CR), meq/100g
A CR of 1 meq/100g	=	1.7 tons gypsum ($CaSO_4$)/acre-ft waste,
and		
A CR of 1 meq/100g	=	0.3 tons elemental sulfur (S)/acre-ft waste
Thus: gypsum requirement	=	CR × 1.7 (tons/acre-ft)
sulfur requirement	=	CR × 0.3 (tons/acre-ft)

We recommend using both gypsum and sulfur when the contaminated materials are sodic, (ESP>15%), and *have solid phase calcium present* to react with sulfur. When using a combination of these materials, the CR is divided by 2 to maintain the equivalent balance and avoid overtreatment.

Relationship of API Gravity To Specific Gravity

Sp. Gr. = 141.5/ (131.5 + ° API)

2

SOIL CHEMISTRY

COLLOIDS — THE REACTIVE SOIL FRACTION

Colloids are in simplest of terms, defined as soil separates smaller than 0.002 mm (2μ) in particle size. They may either be mineral (inorganic) or organic in origin. Because of their small size, soil colloids have a large external surface area. Some layer silicates have an internal surface 50-80 times larger than the external surface and total surface area > 800 m2/g. The most important function of soil colloids is the ability to attract and hold (adsorb) ions of opposite charge on the particle surfaces. For most soil colloids electronegative charges predominate, attracting positively charged ions or cations. The following discusses the source of electronegative charge in mineral and organic colloids. Also discussed is the influence of structural composition on properties of mineral colloids.

Clay Minerals

Clay minerals comprise the most significant group of colloids which is collectively called mineral colloids. Chemical composition and atomic structure in combination define the various clay minerals by group or type.

The atomic lattice of most clay minerals consist of two basic structural units: the silica tetrahedral and aluminum octahedral sheets. They are termed sheets because of the leaf-like or platy stacking arrangement.

Structural Units

The silica tetrahedral sheet is comprised of silicon atoms at the center of and equidistant from four oxygen atoms (Plate 1) arranged to form tetrahedrons (Plate 2) and joined at the base by sharing atoms in the same plane.

The aluminum octahedral sheet is comprised of aluminum atoms at the center of and equidistant from six oxygen or hydroxyl atoms (Plate 3). The octahedral arrangement of aluminum hydroxide $Al(OH)_3$ is demonstrated for gibbsite and kaolinite in Plate 4. Of note is the close order packing of the octahedron units in both minerals and the planar arrangement of hydroxyl and oxygen atoms in the kaolinite structure.

Silicate clay minerals are classified according to the number and arrangement of tetrahedral and octahedral sheets. The 1:1 type minerals consist of one silica tetrahedral sheet and one alumina octahedral sheet. The 2:1 type minerals contain one alumina octahedral sheet sandwiched between two silica tetrahedral sheets.

1:1 Type Clays

Kaolinite is an example of a 1:1 type clay. Chemical and structural features are shown in Plate 5. Attention is given to the shared oxygen atoms holding the tetrahedral and octahedral sheets together. The clay layers are held together by hydrogen bonding, with no room for ions or small molecules between them (Plate 6). There is no interlayer space, which means colloidal properties are determined by external surfaces. Charges at the clay surface arise from broken edges and unsatisfied valences.

2:1 Type Clays

The 2:1 type clays consist of two silica tetrahedral layers and one alumina octahedral layer. The hydrous mica groups, of which illite is a member, carries an electronegative charge due to the substitution of aluminum (Al^{+3} valence cation) for silicon (Al^{+4} valence cation) in the tetrahedral sheet (Plate 7). The illite mineral has electronegative charges satisfied by potassium atoms (K^{+}) which are partially embedded in the tetrahedral sheet between clay layers. This holds the layers together and fixes the interlayer distance. The relatively small, fixed interlayer distance prevents water and cations from entering into the interlayer space (Plate 8). Weathering of the mineral and loss of potassium results in an increase in the effective electronegative charge, exposed interlayers and cation exchange capacity (Fig. 2-1).

The montmorillonite group differs from the hydrous mica group primarily in its expansible nature. In montmorillonite, the source of electronegative charge is the substitution of magnesium (Mg^{+2}) for aluminum (Al^{+3}) in the octahedral sheet (Plate 9). There is little attraction between oxygen atoms in adjacent tetrahedral layers, allowing the movement of water and free exchange of cations into the interlayer space (Plate 10).

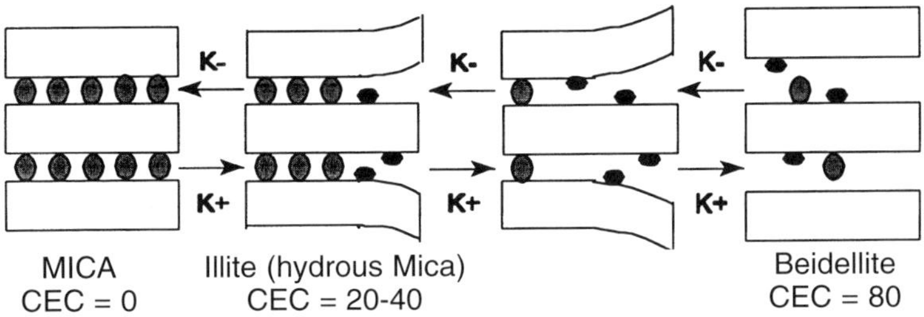

Fig. 2-1 Weathering of the mineral and loss of potassium increases CEC

Mineral Expansibility

Cations like magnesium (Mg^{+2}) and calcium (Ca^{+2}) are hydrated by six water molecules (Plate 11), maintaining an interlayer space of approximately 12 angstroms in montmorillonite. Due to a much larger hydrated radii, the interlayer distance for sodium (Na^{+}) and lithium (Li^{+}) saturated (predominately adsorbed) montmorillonites is not bounded by the cohesive forces exhibited by divalent cations. This may result in particle separation and dispersion. It is this property of montmorillonites that make it useful as a drilling mud. However, this condition must be overcome for successful land treatment.

Sources of CEC

CEC is the total quantity of exchangeable cations that a soil can adsorb, and is generally expressed in milliequivalents of positive charge per 100g of soil (meq/100g). Permanent charge is a term defining the charge generated from isomorphic substitution as demonstrated in the charge balance given in Table 2-1. Other sources of CEC are pH-dependent and may be attributed to hydroxylation and removal of a proton resulting in a net negative charge. Sources of net negative charges are summarized in Table 2-2. Similar functional groups associated with organic molecules also result in pH-dependent sources of negative charges (Table 2-3). Figure 2-2 demonstrates how pH dependence can affect the net charge of various clays.

The cation exchange capacity of soils varies according to pH. These differences are caused by variations in mineralogy, clay content, and organic matter (Fig. 2-3). Table 2-4 shows the CEC of common soil materials, and Table 2-5 illustrates the effect of soil texture on CEC. Generally, soils with high CEC's are more fertile than soils with low CEC's.

Table 2-1 Source of permanent charge cation exchange capacity for Mica, Illite, and Beidellite

Substitution of Al^{3+} for Si^{4+}						
	Unsubstituted (Pyrophyllite)			Substituted Mica, Illite, Beidellite		
	Ions	Charges		Ions	Charges	
		+	-		+	-
	6 O^{-2}		12	6 O^{-2}		12
Silica Layer	4 Si^{+4}	16		4 Si^{+4}	16	
	4 O^{-2}, 2 OH^{-}		10	4 O^{-2}, 2 OH^{-}		10
Alumina Layer	4 Al^{+3}	12		2 Al^{+3}, 2 Mg^{+2}	10	
	4 O^{-2}, 2 OH^{-}		10	4 O^{-2}, 2 OH^{-}		10
Silica Layer	4 Si^{+4}	16		4 Si^{+4}	16	
	6 O^{-2}		12	6 O^{-2}		12
Total		44	44		42	44
Net negative charge			0			2

Table 2-2 Mineral sources of pH-dependent charges

Bonding	Designation	Example	pKa (Acid)
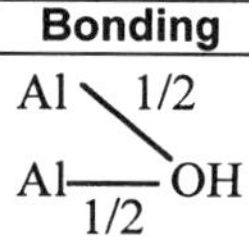	Aluminohydroxyl	$Al(OH)_3$	12+
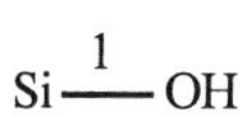	Silica Broken Edge	H_2SiO_3	9.5
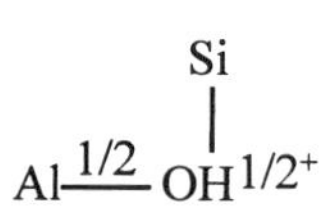	Positive Broken Edge	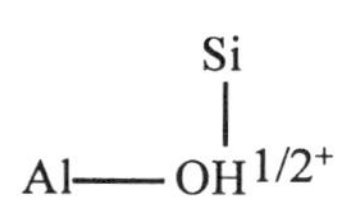	7.0
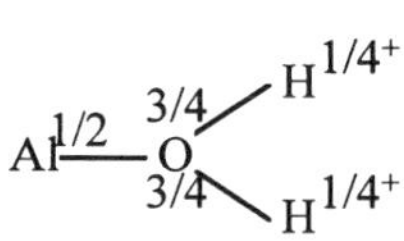	Substituted Hydronium	$Al(OH_2^{1/2+})_6$	5.0

Table 2-3 Organic sources of pH-dependent charges

Functional Group	Structural Formula	pKa (Acid)
Normal Phenol	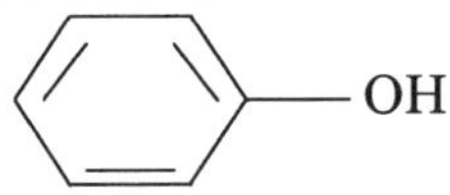	6 - >10
Acid Phenol	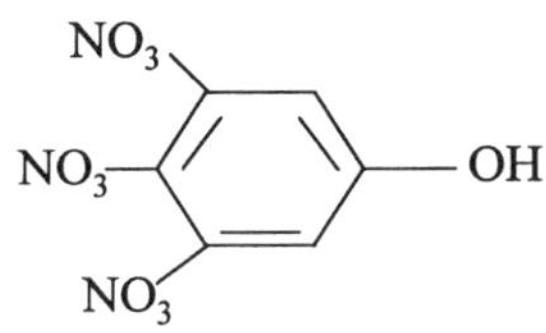	<4
Normal Carboxyl	R—C—OH	4 - >6
Acid Carboxyl	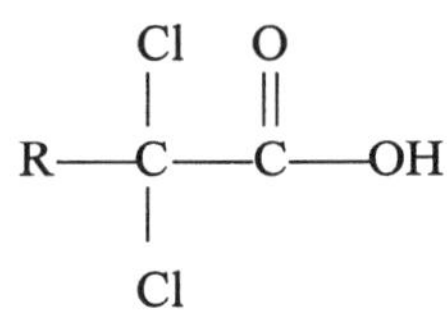	<3

PHYSICAL PROPERTIES OF SOILS

Soil Texture

Soil texture is the size distribution of primary, inorganic particles in soils. The particle sizes are either sand (2 -0.05 mm), silt (0.05 -0.002 mm) or clay(<0.002 mm). The surface area of a clay is 100 to 10,000 times greater than a sand. A loam is a combination of the three particle sizes which exhibit the properties of sand, silt, and clay.

Particle and Bulk Density

Particle density is density of the soil particles. Bulk density is a measure of the weight per unit volume (g/cm^3) of soil or waste. Sandy soils generally have greater bulk density than well-aggregated finer textured soils (Table 2-6). Percent pore space is the space available in the soil for air or water. It is calculated by:

$$\% \text{ Pore Space} = 100 - [(\text{Bulk Density}/\text{Particle Density}) \times 100]$$

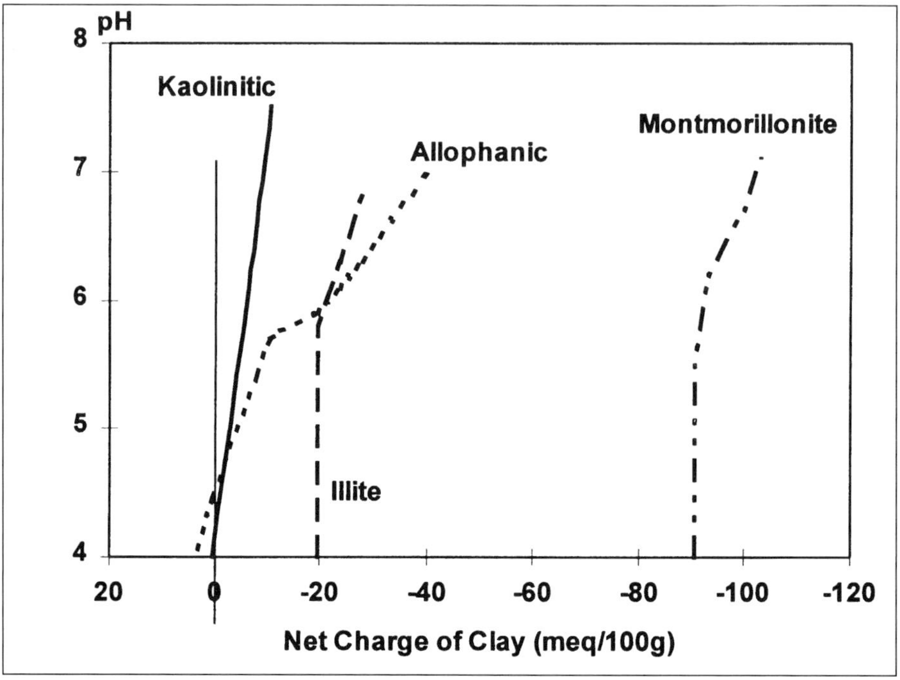

Fig. 2-2 Net charges of selected clays at various pH levels

Table 2-4 CEC of common exchange materials in soils

Material	CEC, meq/100g
Organic Matter	100 - 300
Vermiculite	100 - 150
Montmorillonite	60 - 100
Chlorite	20 - 40
Illite	20 - 40
Kaolinite	2 - 16
Sesquioxides	0

Table 2-5 Relationship between soil texture and CEC

Textural Class	CEC, meq/100g
Sands	1 - 5
Fine sandy loams	5 - 10
Loams, silt loams	5 - 15
Clay loams	15 - 30
Clays	> 30

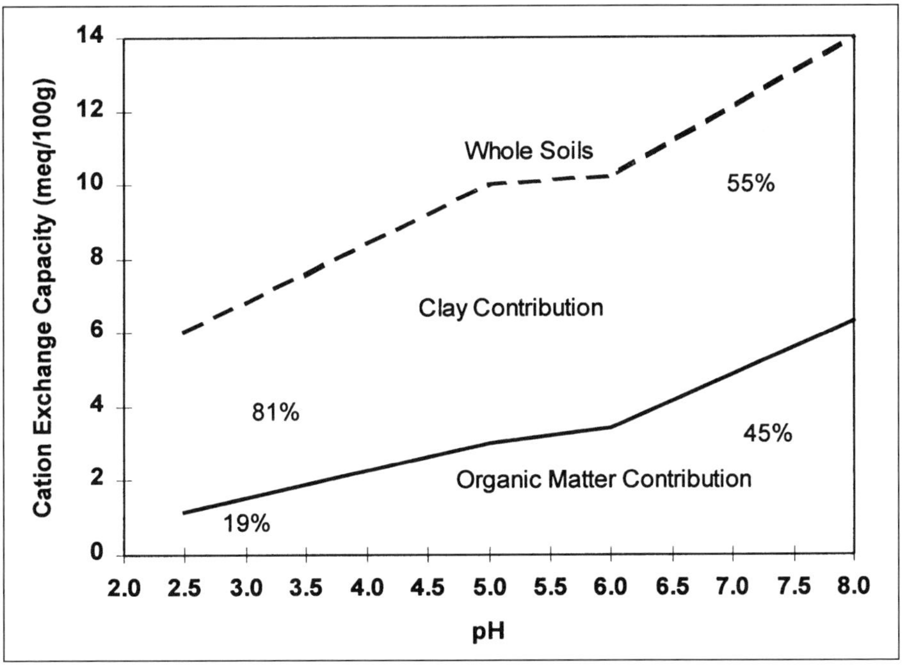

Fig. 2-3 CEC of soil and soil components according to pH

Table 2-6 Soil texture effect on bulk density

Textural Class	Bulk Density, g/cm^3
Sands, Sandy loams	1.2 - 1.8
Silt loam, Clay loam, Clay	1.0 - 1.6

The greater the bulk density of a soil, the lower % of porosity it has (Table 2–7). Macropores are larger pores allowing the ready movement of air and percolation of water. Micropores are small pores in which air movement is impeded and water moves slowly due to capillary rise. A sandy soil has a lower total porosity than clay, but air and water movement is rapid because of dominance of macropores. In clays, water and air movement is slow even though porosity is greater than sands, because of micropore dominance. Water infiltration is faster in a loamy sand than a clay loam. For a soil with a clay loam over sand, the water infiltrates the clay loam layer and does not enter the sand layer until the clay loam layer is saturated due to the severe textural change (Figure 2-4).

Table 2-7 Relationship between soil texture, bulk density, and porosity

Textural Class	Bulk Density, g/cm^3	Porosity, %
Sand	1.55	42
Sandy loam	1.40	48
Fine sand loam	1.30	51
Loam	1.20	55
Silt loam	1.15	56
Clay loam	1.10	59
Clay	1.05	60
Aggregated Clay	1.00	62

Soil aeration is important to keep soil conditions under aerobic processes. Aeration is affected by texture, structure, total porosity, and macropores vs. micropores.

Soil Structure

Soil structure is the way the soil particles group together into stable forms or aggregates. Common types of soil structure are granular, single grain, prismatic, blocky, massive and platy (Fig. 2-5). The structure can

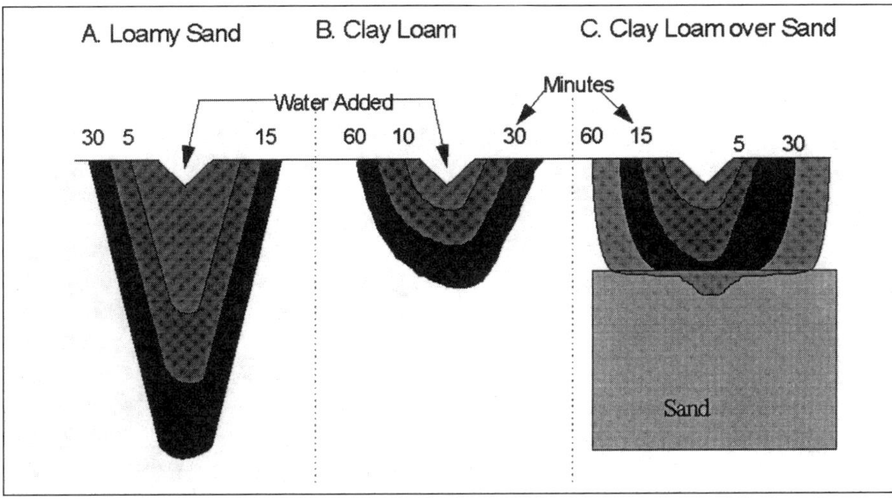

Fig. 2-4 Water infiltration profile for clay loam with and without sand under layer

affect water infiltration. Soils with a granular or single grain can have rapid infiltration, as the water moves quickly through the soil. Blocky and prismatic soils have moderate infiltration. Massive and platy soils have slow infiltration because vertical movement is limited, so the water must move downward through the micropores in the soil.

Soil aggregation and structure can be influenced by many factors. Aggregation tends to increase under conditions of

- Wetting/drying
- Freezing/thawing
- Physical activity of roots and animals
- Decaying organic matter/microbes

Tillage can affect aggregation negatively due to the formation of a hardened plow pan beneath the surface and increased erosion. The effect of adsorbed cations depends on the cations; for example, calcium can improve aggregation, while large amounts of sodium can destroy it. Sodium saturated, structureless soils are often hard packed and require mechanical treatment (chiseling or plowing) to open them up to receive calcium amendments, air and water.

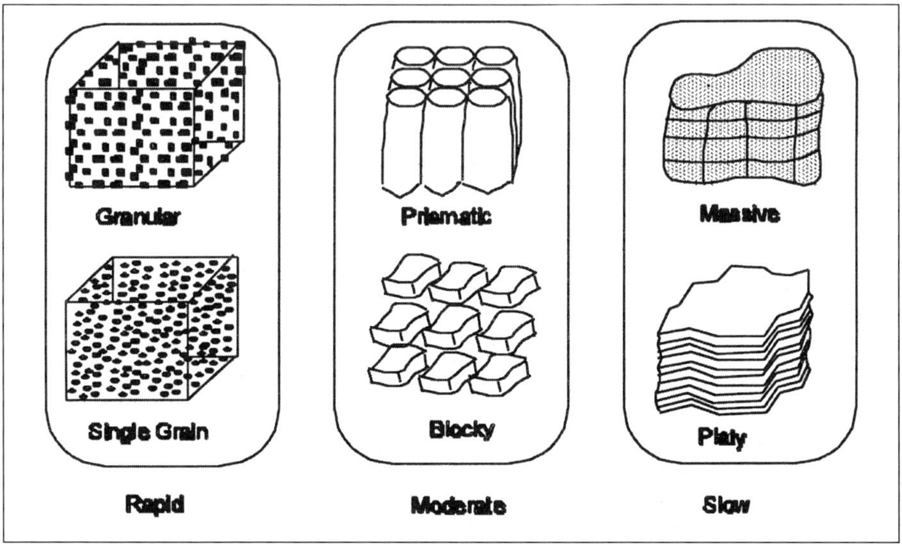

Fig. 2-5 Common soil structures

Organic Matter

Soil organic matter is important for aggregation and structure. Some soil constituents increase aggregation. In order of importance, these are

- microbial gum
- iron oxide
- organic carbon
- clay

Soil organic matter can improve aggregation, water holding capacity, the nutrient reserve, and lower bulk density. As previously demonstrated, a decreased bulk density results in an increased porosity and when combined with improved structure, results in increased macroporosity. This opens the soil up and allows for the infiltration and percolation of rainfall or irrigation water necessary to ameliorate saline and/or sodic soils (See Appendix I for a discussion on the impact of excess salt and sodicity in soils).

Organic matter is an important source of CEC in soils. Adding organic matter increases the receiving soils ability to adsorb cations and therefore serves as a sink for excess salt.

Soil Water Content and Availability

There are several commonly defined levels of soil water content. Field capacity is the water content held in the soil after drainage due to the pull of gravity and matrix effects. The wilting point is defined as the water content at which plants can no longer extract water from the soil. The available water content of a soil is defined as the quantity of water held between the field capacity and the wilting point.

Soil texture has a great effect on both the field capacity and wilting point of a soil. Coarse textured soils (sands and sandy loams) have much lower field capacities and wilting points than fine textured soils (clays), as shown in Table 2-8 and Figure 2-6. The net effect is they contain much less available water and are more susceptible to salt contamination. On the other hand, excess salts are more readily managed by leaching technologies and the addition of organic matter.

Appendix V provides more information on the relationship between field moisture equivalents, saturated soil conditions, and porosity.

Table 2-8 Available water capacities of various soil textures

Textural Class	Water per ft of soil (in)
Medium sand	0.9
Fine sand	1.1
Sandy loam	1.4
Fine sandy loam	1.8
Loam	2.0
Silt loam	2.1
Clay loam	2.0
Clay	1.4

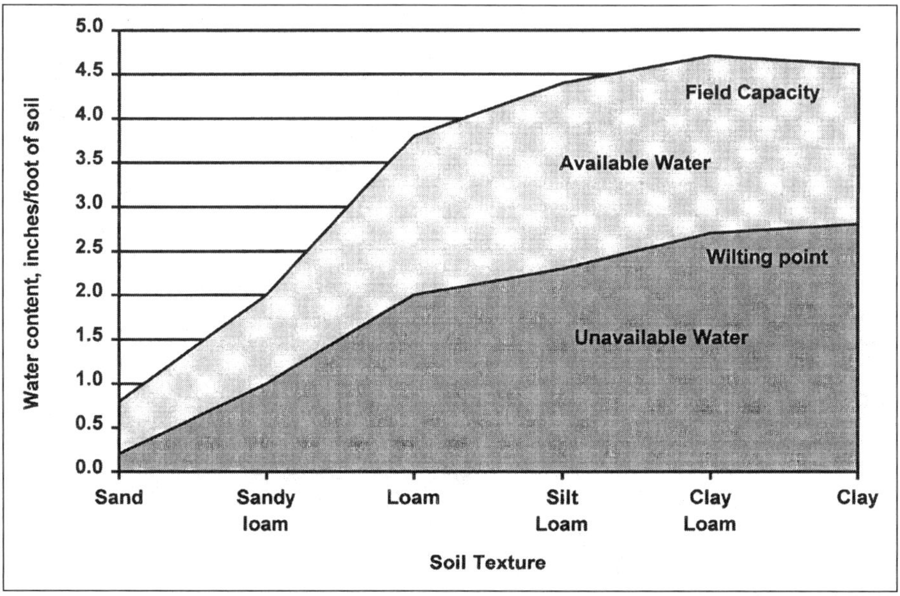

Fig. 2-6 Available water as influenced by soil texture

3

CONTAMINATION AND IMPACTS OF E&P WASTE CONSTITUENTS

Contaminated soils result from (1) drilling operations where drilling fluids become contaminated in the reserve pit during and/or after completion of the well, (2) producing operations where oil, condensate, or produced water are spilled or released onto the ground, and (3) pipeline breaks or leaks where oil and/or produced water are released to the soil.

DRILLING OPERATIONS

Rotary drilling operations result in generating drilling fluids. Drilling fluids can be

1. Fresh water base
2. Salt water base
3. Oil base
4. Emulsion muds

Regardless of the type of muds used, the same contaminants frequently are observed, including changes in pH, Electrical Conductivity, Sodium Adsorption Ratio, Cation Exchange Capacity, Exchangeable Sodium Percentage and total metals.

pH

Drilling fluids typically are alkaline (pH > 10) during use. This high pH disperses the clay and increases the effectiveness of the drilling fluid. The high pH results from the addition of lye, soda ash, or other caustics.

Without the continued maintenance of pH, natural aging and weathering of the drilling fluid causes a decrease of pH through absorption of atmospheric CO_2. The desired range of waste pH is 6-9 standard units (s.u.), as established from discharge criteria from water discharge permits.

The pH determinations for 1,159 E&P reserve pit samples averaged 7.9 s.u. and ranged from pH 1.7 to pH 12.6 (Table 3-1). Cumulative frequency distributions demonstrated 61.4% of the pits fall within the pH 6-9 range, with 13.2% less than pH 6 and 25.4% greater than pH 9.

Table 3-1 Frequency distribution for pH in mud pits

Concentration Interval (su)	Frequency Distribution	% of Total	Cumulative %
0 - 2.0	2	0.2	0.2
2.1 - 4.0	10	0.9	1.1
4.1 - 6.0	140	12.1	13.2
6.1 - 9.0	712	61.4	74.6
9.1 -11.0	242	20.9	95.5
>11.0	53	4.5	100.0
Total	1159		
Maximum	12.6		
Minimum	1.7		
Average	7.9		

Electrical Conductivity (EC)

Electrical conductivity (EC) is the common measure of soil salinity. EC is critical to E&P waste and contaminated soil characterization because of the potential for high brine content and the adverse impact brine has on plant growth and water quality. Produced water and pit liquids are analyzed for EC directly and reported in mmhos/cm. EC is measured for soils and solid wastes using a deionized water extract at a 1:1 (soil:water by weight) or saturated paste (SP) moisture equivalent (Appendix I, Annex A).

Salt distributions measured for 1,198 reserve pits sampled range from EC 0.2 mmhos/cm to 1,559 mmhos/cm and averaged EC = 38.3 mmhos/cm (Table 3-2). Cumulative percentages show problematic salt levels in greater than 59% of pits surveyed. Problematic is defined as materials requiring some management at closure (EC>8.0 mmhos/cm). Acceptable ranges are controlled by post-closure land use as discussed in Chapter 7 and Appendix I.

Table 3-2 Frequency distribution for saturated paste EC in mud pits

Concentration Interval, mmhos/cm	Frequency Distribution	% of Total	Cumulative %
0.0 - 4.0	333	27.8	27.8
4.1 - 8.0	155	12.9	40.7
8.1 - 25.0	349	29.1	69.8
25.1 - 50.0	176	14.7	84.5
50.1 -100.0	77	6.4	90.9
>100.0	108	9.1	100.0
Total	1,198		
Maximum	1,559		
Minimum	0.2		
Average	38.3		

Sodium Adsorption Ratio (SAR)

Soluble cationic distributions are needed to assess the potential sodium damage from wastes. The Sodium Adsorption Ratio (SAR) involves sodium, magnesium and calcium cations and is empirically related to poor soil physical characteristics and plant toxicities. It is defined by a function of sodium, calcium, and magnesium. SAR is computed as follows:

$$SAR = Na / [(Ca + Mg)/2]^{1/2}$$

where cationic concentrations are expressed in milliequivalents per liter (meq/l), and determined for a saturated paste extract. The soil saturated paste moisture content (SP % Moi) is defined as the maximum amount of water held in the puddled soil without free water collecting in a depression made in the soil mass.

SAR is used in conjunction with EC to evaluate the potential damages associated with sodium salts. SAR distributions determined for 1,043 reserve pit samples ranged from SAR less than 0.1 to 2661 with an average SAR of 41.4 (Table 3-3). Greater than 66% of the pit solids analyzed, as well as this average SAR, fail the criteria of SAR<12 suggested in the Limiting Constituents for Land Disposal of Petroleum Extraction Industry Wastes (Table 7-6).

Table 3-3 Frequency distribution for sodium adsorption ratio (SAR) in mud pits

Concentration Interval	Frequency Distribution	% of Total	Cumulative %
0 - 12.0	354	33.9	33.9
12.1 - 14.0	40	3.9	37.8
14.1 - 25.0	149	14.3	52.1
25.1 - 50.0	247	23.7	75.8
50.1- 100.0	158	15.1	90.9
> 100.0	95	9.1	100.0
Total	1,043		
Maximum	2,661		
Minimum	<0.1		
Average	41.4		

Cation Exchange Capacity (CEC)

The individual clay particles in soils and muds have a negatively charged surface on which different positively charged cations are adsorbed. The sum of the positive charges of the adsorbed cations is equal to the sum of the negative surface charge or cation exchange capacity (CEC). Replacement of adsorbed cations by a chemically equivalent number of cations dissolved in soil solution occurs readily. The reaction is reversible depending upon the equilibrium between soluble and adsorbed (exchangeable) cations.

CEC criteria for drilling muds are given as a guide for evaluating acceptable metal loadings. CEC criteria is not a point for consideration when muds contain no appreciable metals. Drilling muds will generally have much larger CEC values than existing surface soils. CEC measurement is required to estimate ESP values. We measure CEC in meq/100g.

CEC distribution for 1,051 pit solids range from a low of less than 1 meq/100g to a high of 84.7 meq/100g and average 20.1 meq/100g (Table 3-4).

Exchangeable Sodium Percentage (ESP)

Just as salt-affected muds are characterized and classified by their content of soluble salts, sodic (excess Na) muds are characterized by the exchangeable sodium percentage (ESP). Saline-sodic drilling muds contain excess salt as well as excess sodium.

When exchangeable sodium is in excess, there is a general lack of structural stability among soil particles, and water infiltration is impeded. Salinity hazards coupled with sodic conditions create soil remediation problems due to the inherent slow infiltration and percolation required to move excess salts out of the root zone.

Table 3-4 Frequency distribution for cation exchange capacity (CEC) in mud pits

Concentration Interval (Meg/100g)	Frequency Distribution	% of Total	Cumulative %
0.0 - 7.0	128	12.2	12.2
7.1 - 14.0	246	23.4	35.6
14.1 - 21.0	235	22.4	58.0
21.1 - 28.0	201	19.1	77.1
28.1 - 35.0	137	13.0	90.1
> 35.0	104	9.9	100.0
Total	1051		
Maximum	84.7		
Minimum	0.8		
Average	20.1		

Sodic and saline-sodic drilling muds in sensitive areas (such as farmland) are not easily managed. Reduced treatment thickness is required for muds having a high (>15%) ESP. Soil amendments, such as sulfur to adjust associated alkalinities and calcium amendments to desorb sodium, are often required in sensitive areas. Sodic hazards are generally indicated for muds with ESP values in excess of 15% (Appendix I).

The frequency distribution for 1,051 pits show ESP values ranging from less than 0.11 to 100, with an average ESP of 28.6% (Table 3-5). More than 60% of the pit solids analyzed fail the criteria of ESP<15 suggested in Chapter 7 and Appendix I.

Total Metals

The chemistry of heavy metals in soil or solid waste matrices is exceedingly complex and as such not completely understood. The ultimate fate of metals contained within a given matrix is controlled by solvation, complexing, chemisorption, and cation exchange processes. These processes control bioavailability and mobility within the receiving soil matrix. Heavy metals are labile in soils.

Metals are distributed in various forms or fractions comprising the solid phase matrix. This includes availability through water soluble, exchangeable, bound- iron/manganese oxides, bound-to carbonates, bound-to organics, and that entrained as part of the mineral fabric. The

Table 3-5 Frequency distribution for exchangeable sodium percentage (ESP) in mud pits

Concentration Interval %	Frequency Distribution	% of Total	Cumulative %
0.0 - 15.0	415	39.5	39.5
15.1 - 25.0	170	16.2	55.7
25.1 - 35.0	148	14.1	69.8
35.1 - 50.0	133	12.6	82.4
50.1 - 75.0	99	9.4	91.8
75.1 -100.0	86	8.2	100.0
Total	1,051		
Maximum	100.0		
Minimum	<0.1		
Average	28.6		

labile pool generally includes the water soluble, exchangeable, and a portion of the organic fraction under moderately well aerated and/or oxidizing conditions. A portion of the heavy metals bound to iron/manganese oxides also may become a significant source of labile metals under poorly aerated and/or reducing conditions.

Water soluble and exchangeable forms define the principal risk associated with heavy metals in the natural environment. If the metal is not in one of these forms, it poses no significant threat or hazard to the environment. Water soluble and exchangeable fractions generally are measured in zero or trace quantities for most E&P wastes and are of no concern in management design. Unfortunately, the natural environment cannot be controlled; redox or pH conditions may change, potentially releasing a measurable and significant fraction of a given heavy metal.

It is our opinion that total metals provide the only measurable index for assessing E&P wastes relative to environmental impact and management design. We recommend E&P wastes routinely be analyzed for total barium, chromium, lead, and zinc. Total arsenic, cadmium, mercury, selenium, and silver should be included when characterizing a field for the first time. However, experience indicates these contaminants seldom are found. Research shows that "total metals released under E&P SW-846, Method 3050 protocol" provides a reasonable index for all metals except barium. Barium is best analyzed under the protocol given in the Louisiana DNR, Office of Conservation Analytical Procedures Manual (Appendix II). The limiting criterion for barium established for drilling muds (20,000 - 40,000 mg/kg)are based on total digestion following Louisiana protocol.

Silver (Ag). Silver is included in the list of total metals typically screened for in E&P wastes because it is one of the RCRA metals. It forms precipitates in soils with chlorides, sulfates, and carbonates and therefore is unlikely to enter the food chain by plant uptake. Silver is allowed in drilling mud pit soil mixtures at levels below 200 ppm. Out of 664 pits tested, all contained silver levels / 10 ppm (Table 3-6).

Table 3-6 Frequency distribution for silver in mud pits

Concentration Interval (mg/kg)	Frequency Distribution	% of Total	Cumulative %
0.0 - 1.0	439	66.1	66.1
1.1 - 3.0	188	28.2	94.3
3.1 - 5.0	29	4.4	98.7
5.1 - 7.0	4	0.6	99.3
7.1 - 9.0	3	0.5	99.8
> 9.0	1	0.2	100.0
Total	664		
Maximum	9.1		
Minimum	<1.0		
Average	1.4		

Arsenic (As). Arsenic naturally occurs in soil in quantities ranging from 1-40 mg/kg. We recommend arsenic be at or below 10 mg/kg for waste-soil mixtures. 86% of pits tested fell under this range (Table 3-7). Under anaerobic conditions the arsenic can be reduced to a form which is toxic to plants. There is still some debate whether or not arsenic is an essential element for plant growth.

Barium (Ba). In soils, the barium content ranges from 100-3,000 ppm. Barite ($BaSO_4$) is commonly used as an additive to the drilling fluid to increase the mud weight. Barite is insoluble in water and the mineral acids used in the EPA total metals digest procedure. Frequency distributions for 1,129 E&P waste pit samples using EPA test protocol showed 92.2% of the pits tested below 8,000 mg/kg (Table 3-8). We find the EPA protocol provides inaccurate and inconsistent results for barium analysis.

Table 3-7 Frequency distribution for arsenic in mud pits

Concentration Interval (mg/kg)	Frequency Distribution	% of Total	Cumulative %
0.0 - 2.0	206	31.0	31.0
2.1 - 5.0	224	33.7	64.7
5.1 - 7.0	71	10.7	75.4
7.1 - 10.0	70	10.6	86.0
10.1 - 20.0	73	11.0	97.0
> 20.0	20	3.0	100.0
Total	664		
Maximum	122.2		
Minimum	<0.1		
Average	5.5		

Table 3-8 Frequency distribution for barium in mud pits

Concentration Interval (mg/kg)	Frequency Distribution	% of Total	Cumulative %
0 - 500	459	40.7	40.7
501 - 1,000	210	18.6	59.3
1,001 - 2,000	115	10.2	69.5
2,001 - 4,000	127	11.2	80.7
4,001 - 8,000	130	11.5	92.2
> 8,000	88	7.8	100.0
Total	1,129		
Maximum	91,100		
Minimum	<10		
Average	2,308		

True Total Barium (Ba). A more accurate level of barium can be measured using a special procedure from the Louisiana 29-B protocol which measures the true total barium. Frequency distribution for true total barium analyzed for 820 E&P waste pits using Louisiana 29-B protocol showed 88.6% of the pits still fall under the wetland regulation of less than 20,000 mg/kg and 93.0% of the pits are acceptable under the upland regulation of 40,000 mg/kg (Table 3-9). We recommend true total barium be maintained at or below 20,000 mg/kg for wetlands and 40,000 mg/kg for upland waste-soil mixtures.

Table 3-9 Frequency distribution for true total barium in mud pits

Concentration Interval (kg/mg)	Frequency Distribution	% of Total	Cumulative %
0 - 5,000	580	70.7	70.7
5,001 - 10,000	86	10.5	81.2
10,001 - 20,000	61	7.4	88.6
20,001 - 30,000	24	2.9	91.5
30,001 - 40,000	12	1.5	93.0
40,001 - 80,000	27	3.3	96.3
> 80,000	30	3.7	100.0
Total	820		
Maximum	270,841		
Minimum	<50		
Average	11,070		

Cadmium (Cd). Depending on the salinity, chemical hardness, and pH, cadmium can be found in reserve pits in the form of CdO, $Cd(OH)_2$ CdS, and $CdCO_3$. Cadmium concentrations as low as 1 ppm can reduce plant growth in certain species. 98.6% of the pits tested fell under the level of 10 mg/kg (Table 3-10). We recommend maintaining cadmium at or below 10 mg/kg in waste-soil mixtures.

Table 3-10 Frequency distribution for cadmium in mud pits

Concentration Interval (kg/mg)	Frequency Distribution	% of Total	Cumulative %
0 - 2.0	594	89.5	89.5
2.1 - 5.0	54	8.0	97.5
5.1 - 10.0	7	1.1	98.6
10.1 - 20.0	6	0.9	99.5
20.1 - 30.0	2	0.3	99.8
> 30.0	1	0.2	100.0
Total	664		
Maximum	37.0		
Minimum	<0.1		
Average	1.2		

Chromium (Cr). Chromium naturally occurring in soils ranges from 5-3,000 mg/kg, with the average level 100 mg/kg. Chromium is not very soluble under the alkaline conditions associated with most drilling muds. Of the 1,129 pits tested, 98.2% of them fell below 500 mg/kg chromium (Table 3-11). We recommend chromium concentrations be maintained at or below 500 mg/kg in waste-soil mixtures.

Table 3-11 Frequency distribution for chromium in mud pits

Concentration Interval (kg/mg)	Frequency Distribution	% of Total	Cumulative %
0 - 50	931	82.5	82.5
51 - 250	163	14.4	96.9
251 - 500	15	1.3	98.2
501 -1,000	7	0.7	98.9
1,001 -2,500	5	0.4	99.3
>2,500	8	0.7	100.0
Total	1,129		
Maximum	11,300		
Minimum	<3		
Average	81		

Mercury (Hg). Mercury naturally occurs in soils in concentrations of <0.2 mg/kg. Mercury often forms stable complexes with clays and organic matter. Only 2 of the 640 pits tested had mercury levels above 10 mg/kg; i.e., 99.7% of the pits were below 10 mg/kg (Table 3-12). We recommend maintaining mercury levels in waste-soil mixtures at or below 10 mg/kg.

Table 3-12 Frequency distribution for mercury in mud pits

Concentration Interval (kg/mg)	Frequency Distribution	% of Total	Cumulative %
0 - 1.0	537	83.9	83.9
1.1 - 2.0	74	11.6	95.5
2.1 - 5.0	22	3.4	98.9
5.1 - 10.0	5	0.8	99.7
> 10.0	2	0.3	100.0
Total	640		
Maximum	23.8		
Minimum	<0.1		
Average	0.7		

Lead (Pb). For native soils, lead concentration is usually about 15 mg/kg. Lead is normally found in insoluble forms in drilling muds. When high chloride contents are present in drilling muds, soluble lead forms may exist. We recommend 500 mg/kg as the maximum lead allowed in waste-soil mixtures. 96.7% of the pits tested fell below this level (Table 3-13).

Table 3-13 Frequency distribution for lead in mud pits

Concentration Interval (kg/mg)	Frequency Distribution	% of Total	Cumulative %
0 - 50	694	61.5	61.5
51 - 250	354	31.3	92.8
251 - 500	44	3.9	96.7
501 - 750	19	1.7	98.4
751 -1,000	5	0.4	98.8
>1,000	13	1.2	100.0
Total	1,129		
Maximum	3,546		
Minimum	<1		
Average	96		

Selenium (Se). Selenium occurs naturally in U.S. soils in the range of 1-7 mg/kg. Selenium usually occurs in the forms of selenate and selenite, which are soluble. Selenium is an essential element for animals, but not for plants. Toxic levels of selenium in the soil can cause stunting and chlorosis in plants. We recommend 10 mg/kg selenium in waste-soil in mixtures. 99.8% of the pits were well below this level (Table 3-14).

Table 3-14 Frequency distribution for selenium in mud pits

Concentration Interval (kg/mg)	Frequency Distribution	% of Total	Cumulative %
0 - 1.0	490	96.1	96.1
1.1 - 4.0	16	3.1	99.2
4.1 - 8.0	2	0.4	99.6
8.1 - 10.0	1	0.2	99.8
10.1 - 15.0	0	0.0	99.8
> 15.0	1	0.2	100.0
Total	510		
Maximum	18.0		
Minimum	<0.1		
Average	0.4		

Zinc (Zn). Zinc levels in soils naturally range from 10-300 mg/kg. Zinc is an essential plant element, needed for the formation of plant hormones and protein synthesis. Zinc is phytotoxic at levels above 600 mg/kg. We recommend an allowable level of 500 mg/kg or less for zinc. 93.0% of the drilling mud pits contain less than 500 ppm zinc (Table 3-15).

Table 3-15 Frequency distribution for zinc in mud pits

Concentration Interval (kg/mg)	Frequency Distribution	% of Total	Cumulative %
0 - 100	663	58.7	58.7
101 - 250	288	25.5	84.2
251 - 500	99	8.8	93.0
501 - 1,000	49	4.3	97.3
1001 - 4,000	28	2.5	99.8
> 4,000	2	0.2	100.0
Total	1,129		
Maximum	6,550		
Minimum	<1		
Average	190		

Petroleum Hydrocarbon

Crude oil from a producing formation and diesel or mineral oil added to drilling muds are the chief sources of petroleum hydrocarbons in E&P wastes. Of these, diesel is the most troublesome with respect to E&P waste management. Fortunately, diesel responds well to bioremediation efforts.

States regulating E&P waste disposal vary in test parameters and applicable treatment thresholds (Chapter 7). Most use either oil and grease (O&G) by gravimetric assay or total petroleum hydrocarbon (TPH) by infrared spectroscopy (IR) for determination of acceptance values. Analytically, the methods yield similar results for unweathered wastes. O&G levels tend to be slightly lower than TPH for fresh wastes in which the lighter hydrocarbons have not been dissipated. Additionally, total organic carbon (TOC) can be used as an alternative analytical method for measuring petroleum hydrocarbons.

Oil & Grease

Oil & Grease determined by solvent extraction and gravimetric analysis for 1,181 E&P waste pits average 5.7% O&G (57,000 mg/kg), ranging from a low of < 0.1% to a high of almost neat oil at 99.1% (Table 3-16). Some 46.2% of the pits in this database meet the Louisiana closure threshold of 1% O&G. Although the data was not split out at the Oklahoma threshold of 2% O&G, it is clear that >50% of the pits would pass this standard prior to any treatment. We recommend a maximum of 2 % TPH.

Table 3-16 Frequency distribution for oil & grease in mud pits

Concentration Interval %	Frequency Distribution	% of Total	Cumulative %
0.0 - 0.5	447	37.8	37.8
0.6 - 1.0	98	8.4	46.2
1.1 - 5.0	272	23.0	69.2
5.1 - 10.0	129	10.9	80.1
10.1 - 20.0	136	11.5	91.6
> 20.0	99	8.4	100.0
Total	1,181		
Maximum	99.1		
Minimum	<0.1		
Average	5.7		

Total Petroleum Hydrocarbon — Infrared (TPH-IR)

Some regulatory bodies require TPH-IR analyses for inventory purposes and comparison to negotiated treatment standards. TPH data for E&P reserve pits receiving only drilling muds is summarized in (Table 3-17). The average was 62,263 mg/kg TPH (6.2%) with a minimum at 13 mg/kg TPH and maximum at 519,336 mg/kg TPH. Cumulative % of pits passing 10,000 mg/kg TPH (1% TPH) was 41.7%, and those passing 50,000 mg/kg TPH (5% TPH) was 65.4%. These values correspond well to the cumulative O&G percentages previously discussed (46.2 passing 1% and 69.2 passing 5%).

Table 3-17 Frequency distribution for TPH-IR in mud pits

Concentration Interval (kg/mg)	Frequency Distribution	% of Total	Cumulative %
0 - 10,000	58	41.7	41.7
10,001 - 25,000	28	20.1	61.8
25,001 - 50,000	5	3.6	65.4
50,001 - 75,000	8	5.8	71.2
75,001 - 100,000	5	3.6	74.8
> 100,000	35	25.2	100.0
Total	139		
Maximum	519,336		
Minimum	13		
Average	62,263		

Total Organic Carbon (TOC)

Petroleum hydrocarbon common to E&P wastes can be analyzed as total organic carbon (TOC) using wet oxidation technique. TOC is recommended as a quick and inexpensive screening technique for evaluating contamination levels for a large number of samples. Also, it can be used as a quality assurance/quality control (QA/QC) measure for either O&G or TPH analyses. TOC frequency distribution for 640 E&P waste pits show average, range, and distribution percentages compare favorably to previously discussed O&G and TPH test results (Table 3-18).

Table 3-18 Frequency distribution for total organic carbon (TOC) in mud pits

Concentration Interval %	Frequency Distribution	% of Total	Cumulative %
0.0 - 1.0	184	28.8	28.8
1.1 - 2.0	104	16.3	45.1
2.1 - 5.0	139	21.6	66.7
5.1 - 10.0	104	16.3	83.0
10.1 - 20.0	84	13.1	96.1
> 20.0	25	3.9	100.0
Total	640		
Maximum	58.0		
Minimum	<0.1		
Average	5.3		

Benzene

Frequency distribution results for total benzene analyzed for E&P wastes (including production sludges) are given in Table 3-19. The average benzene level for 134 separate E&P waste solids was 5.7 mg/kg, ranging from <0.1 mg/kg to 122.9 mg/kg. 82.1% of the waste solids analyzed would pass TCLP benzene threshold of 0.5 mg/kg by calculation (20 × 0.5).

Table 3-19 Frequency distribution for total benzene in E&P wastes

Concentration Interval (kg/mg)	Frequency Distribution	% of Total	Cumulative %
0.0 - 0.5	80	59.7	59.7
0.6 - 5.0	21	15.7	75.4
5.1 - 10.0	9	6.7	82.1
10.1 - 20.0	13	9.7	91.8
20.1 - 50.0	8	6.0	97.8
> 50.0	3	2.2	100.0
Total	134		
Maximum	122.9		
Minimum	<0.5		
Average	5.7		

PRODUCING OPERATIONS

Producing operations typically involve separating oil, water, and gas at a tank battery or satellite station. The separation process occurs in a closed system consisting of heaters, separators, free water knock-outs, and heater treaters. Generally, the only contamination occasioned from these pieces of equipment are due to leaks or purposeful draining of liquids to the soil.

Occasionally, a producing operation includes an emergency pit containing fluids (oil and produced water, primarily) discharged from malfunction of relief valves. These liquids are typically removed immediately after discharge.

The crude oil is stored in atmospheric tanks, generally equipped with equalizing lines. Nevertheless, over-topping the tank does occur. This results in oil saturating the soil. Similarly, produced water may be spilled to the ground. Contaminants of concern include pH, sodium, and hydrocarbons. The same manifestations are experienced in producing operations as with drilling operations. Refer to Drilling Operations for discussion of pH, sodium, and hydrocarbon concentrations.

Produced water pits have not been studied as thoroughly as drilling pits, primarily because operators do not differentiate between pits during remediation. However, one operator identified a group of 51 production pits. The distribution of oil and grease in the solid phase ranged from less than 1.0 to greater than 12% (10,000 mg/kg to 120,000 mg/kg) (Table 3-20).

Table 3-20 Frequency distribution for oil & grease in production pits

Concentration Interval %	Frequency Distribution	% of Total	Cumulative %
0 - 1.0	17	33.3	33.3
1.1 - 3.0	8	15.7	49.0
3.1 - 6.0	7	13.7	62.7
6.1 - 9.0	10	19.7	82.4
9.1 - 12.0	2	3.9	86.3
> 12.0	7	13.7	100.0
Total	51		

Typically, the free hydrocarbons are skimmed from the pit liquids. The liquids normally are injected into a Class II well. As a matter of interest, the produced water pH ranges from 3.4 to 8.7 with an average of 6.8 (Table 3-21).

Table 3-21 pH summary statistics for API produced waters

Test statistic	Produced waters
Sample size (s.u.)	17
Average (s.u.)	6.8
Median (s.u.)	6.9
Mode (s.u.)	6.8
Geometric mean (s.u.)	6.7
Variance (s.u.)	1.2
Standard deviation (s.u.)	1.1
Standard error (s.u.)	0.3
Minimum (s.u.)	3.4
Maximum (s.u.)	8.7

PIPELINE OPERATIONS

The most frequent pipeline releases come from flowlines, which typically contain hydrocarbons and produced water. As wells grow older, the amount of produced water increases. However, the contaminating effects of flowline releases are the same as for other producing operations.

Operators typically install cathodic protection, replace line segments, plastic coat the lines and convert oil tanks to produced water service to reduce line leaks. Nevertheless, increased volumes of produced water enhances the likelihood of releases. Additionally, increased construction along roads and rights-of-ways results in pipeline damage and releases caused by construction equipment accidents. These accidental events are the number one cause of line leaks.

No generalized data are available for pipeline releases since the releases vary so greatly.

IMPACT OF CONTAMINANTS

Heavy metals and sodium adversely impact humans and vegetation, either directly or indirectly. Studies show soluble chloride salts and excess exchangeable sodium are the major ions that migrate in soil. These ions cause harmful effects on the soil and plant growth (Moseley, 1983; Miller & Honarvar, 1975). High soluble salt levels increase the osmotic potential in the soil lowering the amount of water available to plants from the soil. Also, increased osmotic potential interferes with the plant uptake of required nutrients (Murphy and Kehew, 1984; Miller et. al., 1980; Kramer, 1969). High exchangeable sodium levels cause loss of soil structure, resulting in decreases in water and air infiltration and root zone soil compaction (Moseley, 1983; Miller et. al., 1980).

Soils having high organic matter content and moderate percentages of clay appear less sensitive to high sodium and soluble salt concentrations than coarser textured soils or loamy or clayey soils in a dry environment (Moseley, 1983; Deuel, 1991; Miller, 1975).

Heavy metals found in drilling muds include chromium, cadmium, and lead. There is concern these metals might become incorporated and accumulated in the food chain or contaminate local groundwater if leaching occurs from drilling pits. The availability (mobility) of metals in soil is dependent upon pH, reduction/oxidation potential and total metals concentration. Matrix pH is the controlling variable. In general, solubility of metals is inversely related to pH, except for barium, whose solubility is directly related to pH (Freeman and Deuel, 1984.)

Migration of metal ions away from a pit site is usually limited by their attenuation in clay minerals in the soil. Also, metal ions tend to form insoluble complexes in soil, especially at high pH levels (Murphy and Kehew, 1984; Moseley, 1983). Researchers find little or no migration of metal ions from drilling muds because of clay attenuation and complexing (Whitmore, 1982; Henderson, 1982). Attenuation and migration are affected by soil type and proximity of ground water. Migration is much more extensive in porous soils than in clayey soils and where ground water is close to the pit bottom (Murphy and Kehew, 1984).

Deuel and Holliday (1997) conducted a laboratory study of geochemical forms of heavy metals in E&P wastes to evaluate their potential bioavailability and mobility in the environment. They concluded low total and/or low exchangeable metal concentrations in E&P wastes control the bioavailability and mobility of heavy metals in soil.

References

Deuel, L.E. 1991. "Evaluation of Limiting Constituents Suggested for Land Disposal of Exploration and Production Wastes." API, 1220 L Street, Northwest, Washington, D.C. 20005.

Deuel, L. E. and Holliday. 1997. "Geochemical Partitioning of Metals in Spent Drilling Fluids." Presented at ASME Energy Week, Houston, TX, January 28-30.

Freeman, B.D. and L.E. Deuel. 1984. "Guidelines for Closing Drilling Waste Fluid Pits in Wetland and Upland Areas." Presented at the 7th Annual Energy Sources Technology Conference and Exhibition. Sponsored by The Petroleum Division - ASME, New Orleans, LA, February 12-16.

Henderson, G. 1982."Analysis of Hydrologic and Environmental Effects of Drilling Mud Pits and Produced Water Impoundments." Dames & Moore. API Production Department, Dallas, TX.

Kramer, P.J. 1969. Plant and Soil Water Relationships: A Modern Synthesis. McGraw-Hill Inc., New York.

Miller, R.W. and S. Honarvar. 1975. "Effect of Drilling Fluid Component Mixtures on Plants and Soil". Environmental Aspects of Chemical Use in Well Drilling Operation, Conference proceedings, May, Houston, TX. EPA 560/1-75-004. pp. 125-143.

Miller, R.W., S. Honarvar, and B. Hunsaker. 1980. "Effects of Drilling Fluids on Soils and Plants: I. Individual Fluid Components." J. Environ. Qual. pp. 9:547-552.

Moseley, H.R. 1983. "Summary and Analysis of API Onshore Drilling Mud and Produced Water Environmental Studies." American Petroleum Institute Bulletin D 19.

Murphey, E.C. and A.E. Kehew. 1984. "The Effect of Oil and Gas Well Drilling Fluids on Shallow Groundwater in Western North Dakota." Report No. 82. North Dakota Geological Survey.

Rule 29-B. 1990. Section XV - Pollution Control, State of Louisiana, Department of Natural Resources, Office of Conservation, Baton Rouge, LA, Promulgated January 20, 1986, revised October 20, 1990.

Whitmore, J. 1982. "Water Base Drilling Mud Land Spreading and Use as a Site Reclamation and Revegetation Medium". Fosgren-Perkins Engineering. API Production Department, Dallas, TX.

4

SAMPLING METHODS AND PROTOCOLS

SITE INVESTIGATION

Site investigation procedures require a plan of action to maximize the safety of sampling personnel, minimize sampling time and cost, reduce errors in sampling, and protect the integrity of the samples after sampling. Proper sampling procedures are:

1. SELECT proper sampling device
2. SELECT proper sample container and closure
3. DESIGN sampling plan to include:
   a) SAMPLING POINTS
   b) SAMPLE QUANTITIES
   c) SAMPLE COMPOSITING, if desired
4. IDENTIFY and PROTECT samples from tampering
5. RECORD sample information in field note book
6. Complete CHAIN of CUSTODY record
7. Complete SAMPLE ANALYSIS request sheet
8. DELIVER/SHIP samples to laboratory for analysis

GENERAL GUIDELINES

Before collecting a sample, decide what components are to be analyzed and why. For example, many exploration and production (E&P) operations generate multiphasic matrices consisting of both liquid and solid phases. Also, liquids may be multiphasic with a free floating oil layer over water.

Composite samples should reflect all phases in proportion to their occurrence. Phases to be treated, discarded and/or disposed of separately should also be sampled separately.

Sample Containers

Collect samples in glass containers fitted with plastic or metal screw lids. One quart or one pint canning jars work well for all inorganic parameters. For organic parameters such as benzene or TPH, glass jars with Teflon®, lined lids are recommended. Each jar should be wrapped tightly with cellular foam materials to reduce breakage from jars colliding during shipment. Also, each jar should be placed in a sealable storage bag to avoid sample loss in event of jar breakage. Alternatively, soils or pit solids can be collected into sealable storage bags. However, extreme care must be exercised to avoid contamination of the locking device which could cause sample leakage.

Samples collected for "permitted discharge" or analysis by NPDES criteria or USEPA hazardous waste characterization should be sampled in containers preserved and held per Table 4-1.

Table 4-1 Water sample holding times, required digestion volumes and recommended collection volumes for metal determinations

Measurement	**Digestion Volume Required (ml)**[a]	**Collection Volume (ml)**[b]	**Preservative**	**Holding Time**
METALS (Except hexavalent chromium and mercury)				
Total Recoverable	100	600	HNO3 to pH<2	6 mo
Dissolved	100	600	Filter on site HNO3 to pH<2	6 mo
Suspended	100	600	Filter on site	6 mo
Total	100	600	HNO3 to pH<2	6 mo
Chromium VI	100	400	Cool, 4∞C	24 hr
Total Mercury	100	400	HNO3 to pH<2	28 days
Dissolved Mercury	100	400	Filter on site HNO3 to pH<2	28 days

[a] Solid samples should be at least 200g and usually require no preservation other than storing at 4°[1] C until analyzed.

[b] Either plastic or glass containers may be used.

Sample Labeling

Label all samples immediately after placing the sample into the container. Use of a pre-printed self adhesive label facilitates labeling. DO NOT write directly on the glass; permanent ink will smear if the glass gets wet. Each label should contain a complete sample description, including date, time, sample location, description, and name of sampling personnel. Confusion is avoided if the label includes the analytical parameters to be performed by the laboratory. A chain of custody form or laboratory log-in sheet can be used to track the sample during sampling, delivery and analysis. Any sample used during legal procedures must have an accompanying chain of custody sheet signed by the sampler, transporter, receiver, and laboratory personnel.

Sample Preservation

Sample storage and holding times need to be strictly followed. Samples collected for Louisiana Statewide Rule 29-B analyses must be preserved on ice. Samples collected for "permitted discharge" or analysis by NPDES criteria or USEPA hazardous waste characterization should be held according with EPA times (Table 4-1). Holding samples beyond the recommended time subjects the analytical results to question.

EQUIPMENT DECONTAMINATION

All field sampling equipment should be laboratory cleaned, wrapped, and dedicated to a particular sampling point. If this is not possible, sampling equipment may be cleaned in the field, using the following cleaning strategy:

1. Non-phosphate detergent/tap water wash
2. Tap water rinse
3. Distilled/deionized water rinse
4. 10% nitric acid rinse* (trace metal or higher grade nitric acid diluted with distilled/deionized water)
5. Distilled/deionized water rinse*
6. Acetone (pesticide grade)**
7. Total air dry**
8. Distilled/deionized water rinse**

   * for metals/inorganic analysis
   ** for organic analysis and TPH

SAMPLE COLLECTION

Sampling Water

For samples to be taken from a valve, spigot, or flow line, the sampling point is to be flushed a minute or two before actual collection. Fill the sample container almost full. DO NOT OVERFILL. The exception is when sampling for volatile organics (i.e. benzene, toluene, ethylbenzene, xylenes, etc.) which require zero headspace for preservation of samples. In this case, samples are collected by overfilling containers prior to sealing.

To sample water from an auger hole or a well a bailer may be used. Insert the bailer into the hole until it reaches water. Allow the bailer to fill with water and collect the samples.

Sampling Solids

Many different devices can be used to take soil samples. There are different types of augers which are used depending on the soil type. For sandy soils, a sand auger should be used (Fig. 4-1). The teeth on the cutting end of the sand auger intersect so the sand doesn't fall out when the sample is

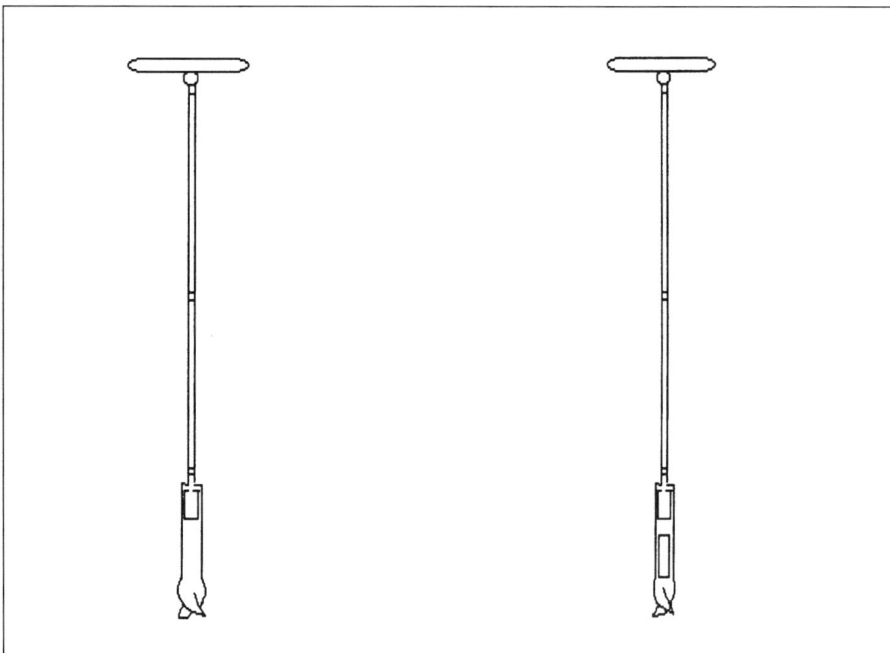

Fig. 4-1. A sand auger is depicted on the left and a mud auger on the right

pulled out of the ground. For clays or drilling muds, a mud auger with open sides works best (Fig. 4-1).

A soil probe works well when the sample depth is less than 3 feet. Soil probes do not work well in sandy or compact soils. The sample size is very small with a soil probe (typically 3/4 inch in diameter); therefore, many samples must be taken and then composited.

To take undisturbed soil cores, a split core sampler is needed (Fig. 4-2). Often a hammer attachment is needed to drill the split core into the ground. Disposable liners may be used and sent to the lab as collected.

Sampling Pit Sludges

Sampling of sludges or sediments from storage pits, impoundments, ponds, or ditches usually can be achieved by pushing a hollow tube (metal or PVC plastic, 2 in internal diameter) into the pit bottom through all layers of waste, so the sample is representative of the waste matrix and impacted soil or sediment (Fig. 4-3). Unconsolidated samples are retrieved by pushing the tube into consolidated native soil to form a plug holding the

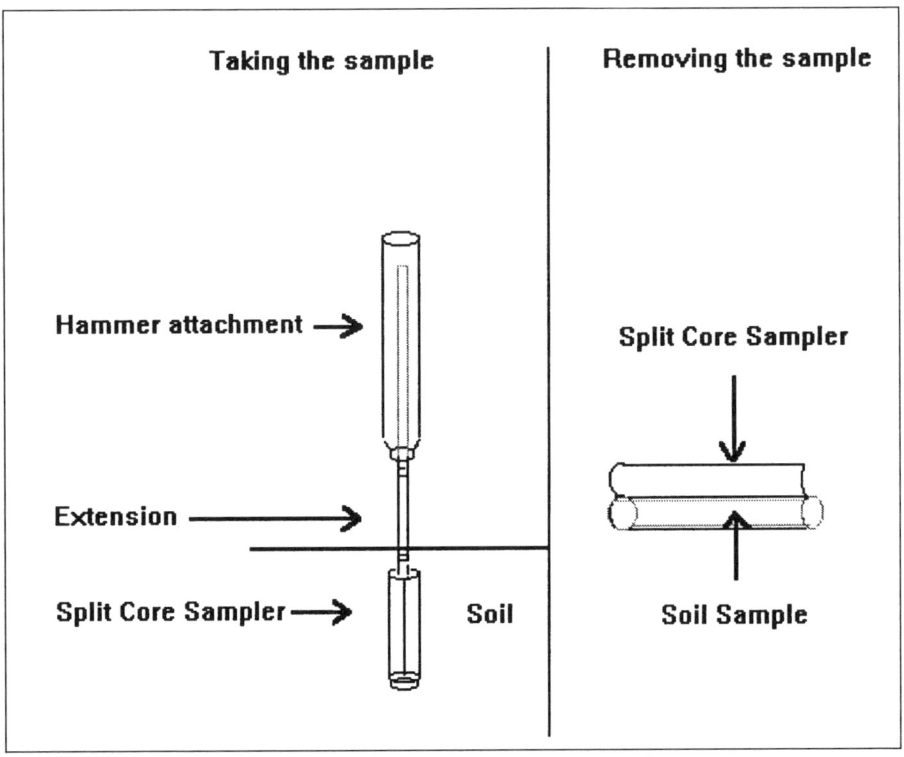

Fig. 4-2. Taking a sample using a split core sample

sample in the tube. DO NOT include the native soil plug as part of the sample. An end cap or plugging device, such as a rubber stopper or your hand, can be placed on the top end of the tube after sampling, creating a suction to keep the sample from discharging upon retrieval. Some solids may have to be discharged to the sample container with the aid of a smaller diameter ramrod. A flat-bottom boat may be needed to obtain access to all sampling points.

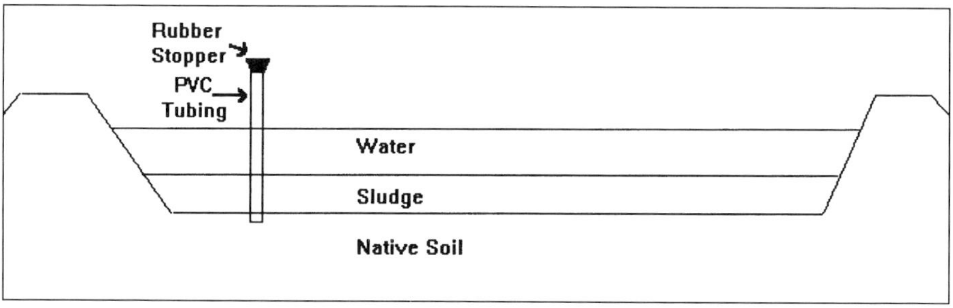

Fig. 4-3 Representative sampling of pit wastes using PVC tubing

Surface or shallow samples can be taken with a scoop or shovel provided all layers are included in correct proportion to the sample matrix. All cores and/or layers within cores to be composited are blended in the field prior to taking the representative sample.

Sampling Solids Covered With Water

Occasionally, it is necessary to sample solids located beneath water, such as a river or lake bottom. Care must be taken to sample the solid matrix only, without sampling the water. One method is to take a 6 inch diameter PVC tube, sealed at the lower end with plastic wrap and rubber band (Fig. 4-4). Insert the tubing into the solids beneath the water at the sample location approximately 2 inches deep. Insert an auger into the PVC tubing and puncture the plastic wrap. Use the auger to take the solid samples needed; being careful to keep water from entering the tubing.

COMPOSITING SAMPLES

Samples should be composited on a weight/weight or volume/volume basis under controlled conditions. Compositing should be limited when analyzing for volatile organic compounds because compositing caus-

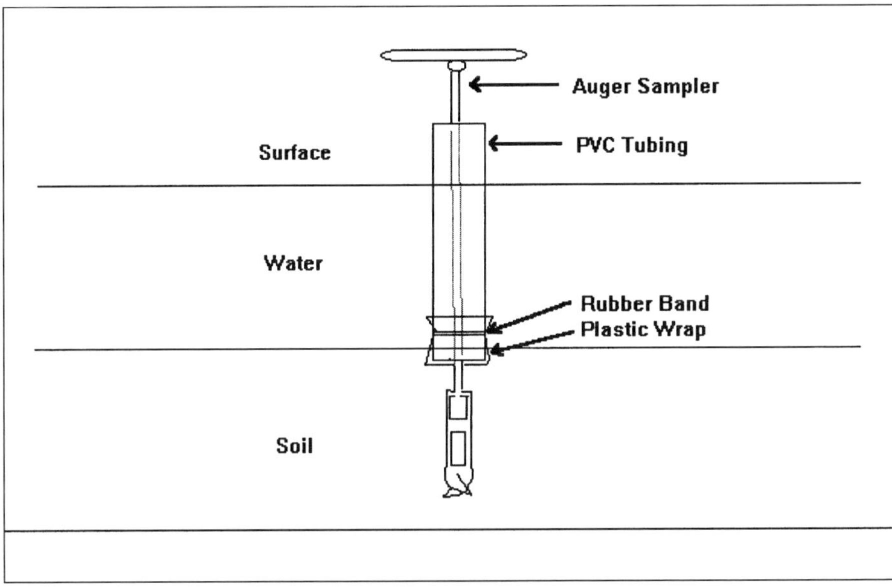

Fig. 4-4 Sampling solids located beneath water without contaminating the solids

es a partial loss of volatile constituents. After careful and complete homogenization of each individual solid sample, weigh equal portions of each into one container. After the composite is made, homogenize the composite completely before analyzing.

Liquid samples should be mixed well by shaking or stirring before compositing. Equal portions (measured as either volume or weight) are combined into one container. All phases of multi-phase samples should be represented in the composite in the proper proportions.

FIELD QUALITY ASSURANCE

Proper field QA procedures include full documentation of sampling methods in a field notebook, use of trip or travel blanks, and generation of blind split samples in the field.

Field Notebook

Full documentation using a field notebook is required for all samples.

Information should include date and time samples were taken, surface location, depth, sample matrix, soil type, and evidence of contamination (odor, color). Care should be taken to clearly identify sample containers.

Trip Blank

As a check for improperly cleaned sampling containers or trip contamination, a sample of each type of container that will be used in the field on a given day should be filled in the laboratory with distilled water, sealed, and transported to and from the field with other sampling containers. The trip blank should then be analyzed alongside field samples.

Split Samples

Approximately 10% of samples to be analyzed (typically after homogenizing and compositing), should be split and placed into separate containers. The samples should not be labeled so as to be identifiable as split samples by laboratory personnel. Although with soil and sludge samples, some variability can be expected between split samples, reasonable correlation is expected between samples.

References

Rule 29-B. 1990. Section XV - Pollution Control, State of Louisiana, Department of Natural Resources, Office of Conservation, Baton Rouge, LA, Promulgated January 20, 1986, revised October 20.

5

SAMPLING PROGRAM

The pit sampling program is designed identically for large and small impoundments. Typically, the large pit is divided into areas. Each area is sampled independently of the other area.

Think of small pits as one or more areas of a larger pit and sample accordingly.

LARGE FACILITY IMPOUNDMENTS

Large facility impoundments typically receive the full spectrum of E&P wastes, which can be broadly classified into 3 major groups:

1. Weathered drilling fluids and drill cuttings from past operations
2. Recent spent drilling wastes reflecting newer formulations and/or
3. Tank bottoms, and other produced liquids and sludges

Impoundments source variability, size, and natural segregation of materials with depth and distance from the point of discharge, presents a challenge for obtaining representative samples. Representative samples are fundamental for characterization of pit content parameter variability, which in turn is necessary in establishing the scientific validity and relevance of the analytical measurement.

Statistical Validation

A large commercial pit was selected as a means to verify waste distribution within the pit. The pit was divided into 15 areas (Figure 5-1). Each area was subdivided into 4 equal sampling sites (Figure 5-1). Random duplicate sample cores were taken from a flat bottom boat to native substrate or 3 meters (whichever came first).

Each sample was placed into a glass container and labeled as to area, sampling site and replicate. Each sample was homogenized and analyzed for pH.

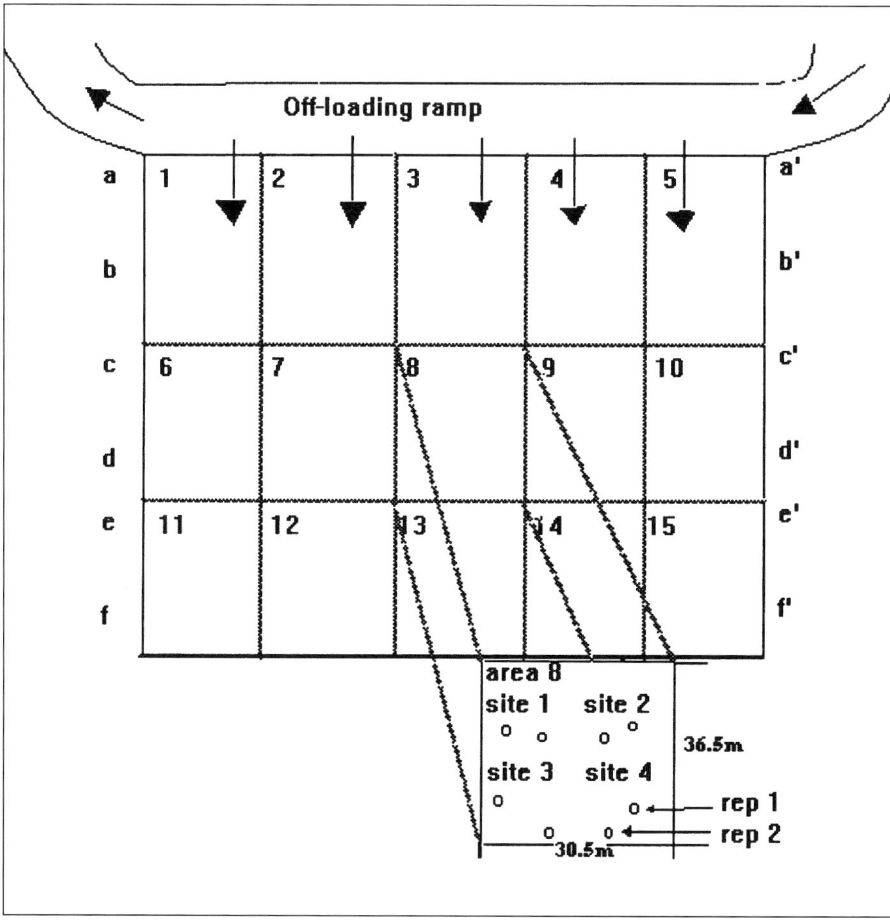

Fig. 5-1 Schematic showing pit delineated by area, sites within areas, and replicate core locations.

An analysis of variance (ANOVA) of pH showed pit material pH varied significantly between areas and between sampling sites (Table 5-1). The F-test shows a significant interaction between areas and sampling sites (Table 5-1). The analysis showed no significant interaction between replicate samples.

Table 5-1 The ANOVA for pH in large pit samples

Source	df	SS	Mean square	F
Treatments				
reps	1	0.0078	0.0078	0.067
area (a)	14	76.5699	5.4693	47.009*
site (s)	3	13.9694	4.4693	43.023*
a X s	42	39.6021	0.9429	8.104*
Error	59	6.8644	0.1163	
Total	119	137.0138		

*denotes highly significant difference, $p < 0.01$.

Table 5-1 shows that pH values averaged over each replicate sample by sampling site zones show much higher values at the offloading ramp. Also, Table 5-2 shows the results of a Tukey multiple range test comparing mean pH values between sampling sites and zones.

Results show unit 1 yielding the highest pH (Table 5-2), followed by unit 2, which was significantly greater than units 3 and 5. Unit 4 was significantly lower in pH than the aforementioned units but was statistically equivalent to all others. Site means averaged over reps and units revealed sites 1 and 2 significantly higher in pH than sites 3 and 4.

A three-dimensional surface plot shows the interaction between area and site and the effect of this relationship on pH (Figure 5-2). The fact inner units are not significantly different suggests that the large E&P facilities impoundments could be sampled close to the sides. One should sample a sufficient distance from the bank such that the depth of the materials encountered and the potential for layering is representative of the pit as a whole.

Figure 5-2 graphically displays pH highest near the offloading ramp (units 1-5) and decreases with distance away from the ramp. Apparently, the trucks pull to the northwest corner of the pit during unloading since the pH is highest in Areas 1 and 2. Further, when several trucks unload, the second truck appears to unload near the northeast corner of the pit. We conclude the distance from the point of discharge has a significant effect on waste composition.

Table 5-2 Mean pH values by unit and sites

			Site		
Unit	1	2	3	4	Average
1	11.17	10.73	10.55	8.39	10.21a
2	11.00	11.55	7.68	7.64	9.47b
3	9.71	9.52	7.87	7.51	8.65c
4	7.81	7.82	7.53	7.48	7.66d
5	8.12	10.03	7.57	8.09	8.45c
6	8.53	7.68	7.29	7.27	7.69d
7	7.81	7.71	7.97	7.62	7.78d
8	7.49	7.64	7.45	7.49	7.52d
9	7.52	7.23	7.55	7.43	7.43d
10	7.40	7.64	7.50	7.39	7.48d
11	7.78	7.85	7.62	7.84	7.76d
12	7.98	7.79	7.52	7.46	7.69d
13	7.52	7.75	7.44	7.56	7.57d
14	7.66	7.59	7.46	7.55	7.56d
15	7.39	7.43	7.59	7.46	7.47d
Average	8.32a	8.40a	7.77b	7.61b	

Averages not followed by the same letter are not significantly different at the 5 percent level.

Zonal Composites

Transect composite samples were prepared by combining sub-samples of each core taken along a transect. Zonal composites 1,2, and 3 were prepared by combining subsamples of transect composites a-a′ and b-b′; c-c′ and d-d′; and e′ and f-f′ respectively (Figure 5-1). The zone 1 composite is equivalent to combining all rep subsamples from units 1 through 5. Zone 2 represents units 6 through 10, and zone 3 represents units 11 through 15.

Zonal composites were analyzed for the suite of parameters listed in Table 5-3 to better evaluate materials composition and distribution relative to the point of discharge. The moisture content of solids was highest in zone 3 materials, demonstrating that the finer dispersed particles tend to be segregated with distance from the point of discharge. This was reflected in the higher cation exchange capacity (CEC) and greater exchangeable sodium percentage (ESP) of the zone 3 solids. Soluble phase constituents including electrical conductivity (EC), and cationic speciation were approximately equal in all three zones. Total petroleum hydrocarbon (TPH) was also

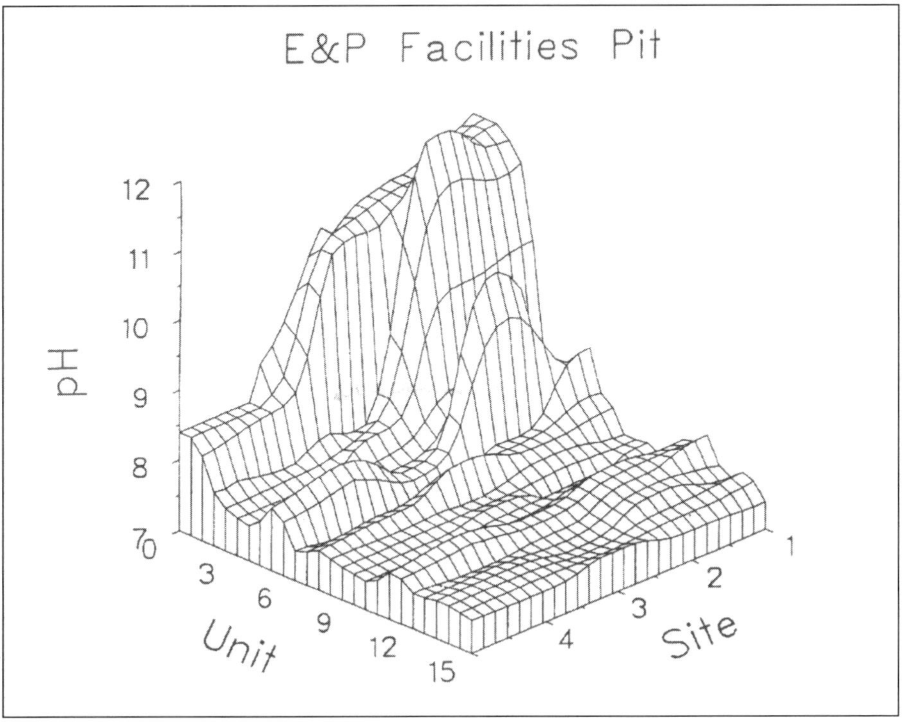

Fig. 5-2 Surface plot depicting the area site interaction

approximately equally distributed. Total metal compositions varied considerably between zones and apparently between sources. Barium, for example, was highest in zone 1, while chromium was highest in zone 3. This suggests that barium is related more to the coarser fraction of the waste as an additive, and chromium exists more as a tramp contaminant of clay.

Future Sampling Considerations

Based on the materials composition and distribution of solids in the large impoundments, it is clear that a minimum of two samples should be generated for each pit.

One approach to collecting the two samples is given in Figure 5-3. An aluminum boat or even a crane with a basket platform can be used to collect cores. The main objective is to get out far enough into the pit that a vertical profile can be taken that is representative of the waste solids stratification.

Table 5-3 Zonal composite analyses

Parameter*	Zone 1	Zone 2	Zone 3
Moisture, %	73.7	152.6	211.6
pH, s.u.	9.7	8.6	9.7
SP EC, mmhos/cm	11.1	9.4	10.1
Soluble Cations, meq/liter			
Sodium	62.5	59.9	60.5
Calcium	11.4	4.0	9.9
Magnesium	<0.1	0.5	<0.1
Potassium	<0.1	<0.1	<0.1
CEC, meq/100g	22.8	31.1	35.7
Exchangeable Cations, meq/100g			
Sodium	1.1	4.9	8.3
Calcium	2.2	4.8	4.8
Magnesium	0.6	2.2	2.4
Potassium	<0.1	<0.1	<0.1
ESP, %	4.8	15.6	23.3
Total Metals, ppm			
Barium	2,326	1,307	1,051
True Total Barium	179,816	103,030	87,864
Chromium	74.0	126.6	190.4
Lead	62.2	59.1	61.8
Mercury	0.1	0.1	0.3
Zinc	240.9	211.1	207.1
TPH, %	5.6	4.2	5.3
TOC, %	3.8	4.3	3.7

*Analysis methods are from Laboratory Procedures for Analysis Oilfield Waste, Louisiana DNR, August, 1988.

Duplicate cores should be taken from 5 to 7 sampling points and blended together to form zone 1 and zone 2 composites. Pit characterization should include analysis of both composites.

FIELD OR RESERVE PITS

Field pits are identified as earthen impoundments used to receive E&P wastes during the drilling process. They are much smaller than the facility

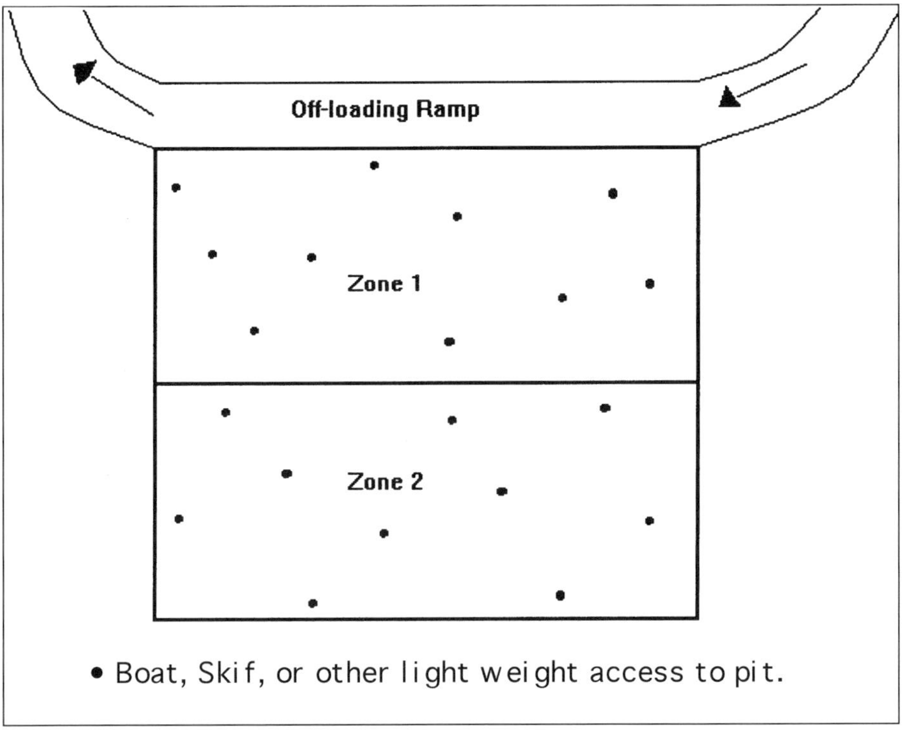

Fig. 5-3 Schematic showing recommended sampling points

impoundments and therefore more easily sampled in representative fashion.

Sampling Strategy

An example of a field pit used operationally as a reserve mud pit is shown in Figure 5-4. Because of its size this particular pit was subdivided into 6 sections. Sections were accessed by a flat bottom boat. Solid samples were collected from 10 sites within each section by pushing a hollow tube through to consolidated native soil defining the pit bottom(Figure 4-3). Site cores were combined yielding a representative section sample. Section samples were subsampled individually into glass containers prior to blending into 1 pit composite.

Core Composite Validation

Several metal parameters were analyzed for individual section samples to determine their distribution in the reserve pit relative to the point of discharge (Table 5-4). All metals except sodium were highest in concentration in those sections nearest the points of discharge (sections 1, 2 and 4).

Sodium was distributed in reverse to this trend with concentrations highest in sections 5 and 6. Obviously, sodium reflects the dispersed nature of the clay gels used in drilling operations, with the fine particulates tending to remain in suspension at a distance from the point of discharge. Calcium saturated clays tend to be flocculated and fall out with coarser clay particulates and other solid additives.

Table 5-4 Metal parameter analyses by area (mg/kg)

	Area		
Metal Parameter	1	2	3
Barium	38,507	39,761	12,000
Chromium	34	20	15
Lead	60	56	35
Zinc	184	230	87
Arsenic	6.6	5	2.9
Iron	21,560	15,976	10,782
Manganese	391	202	191
Sodium	1,600	2,800	2,680
Calcium	38,600	20,000	9,600
Magnesium	4,290	3,040	2,500
Potassium	3,923	3,510	3,132
	Area		
Metal Parameter	4	5	6
Barium	26,992	6,317	6,143
Chromium	26	24	29
Lead	52	32	37
Zinc	32	12.4	13
Arsenic	6.4	4.9	4.8
Iron	20,946	17,725	16,555
Manganese	221	213	209
Sodium	4,920	5,660	5,030
Calcium	22,800	15,100	12,200
Magnesium	4,310	4,110	4,140
Potassium	4,220	4,135	4,117

*Analysis methods are from Laboratory Procedures for Analysis of Oilfield Waste, Louisiana DNR, Aug., 1988

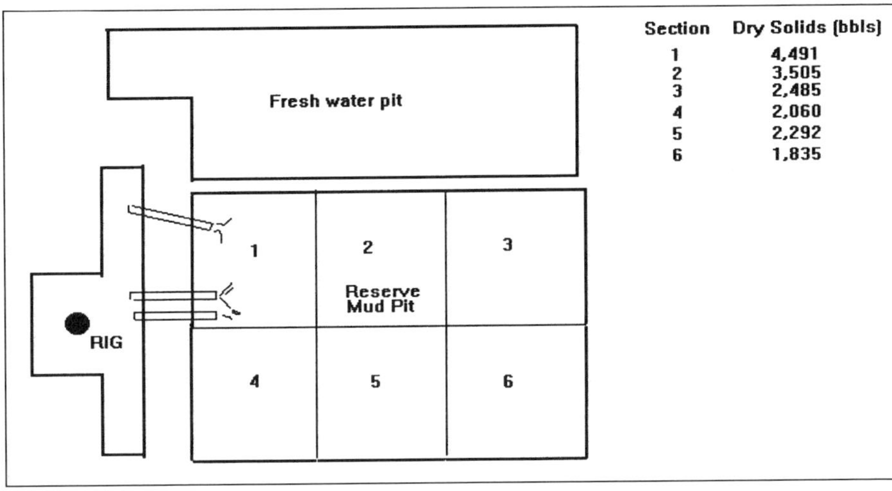

Fig. 5-4. Schematic of field (reserve) pit in operation

Weighted averages calculated from individual section sample analyses are in close agreement with the overall pit composite (Table 5-5). The data suggests that multiple core composites provide an acceptable means for obtaining representative samples from both areal and spatially variable matrices. Composites also provide a means for reducing analytical loads and associated costs.

Table 5-5 Comparative pit solids analysis

Metal Parameter	Section Average mg/kg	Pit Composite mg/kg
Barium	25,406	29,191
Chromium	25.3	16
Lead	47	64
Zinc	118	120
Arsenic	5.3	3.1
Iron	17,624	15,100
Manganese	256	289
Sodium	3,360	3,450
Calcium	22,275	26,980
Magnesium	3,722	4,110
Potassium	3,375	2,995

*Analysis methods are from Laboratory Procedures for Analysis of Oilfield Waste, Louisiana DNR, August., 1988

Future Sampling Considerations

In general, field pits should be subdivided into approximately 30 m^2 areas sections, with each unit represented by 5 to 7 core composites. A 60 m^2 pit yields two unit samples; a 90 m^2 pit generates 3 unit samples; etc. Unit samples are transported to the laboratory in separate containers so that pit composites can be prepared in proportion to their occurrence. Typically, pit composites are generally analyzed for the full suite of parameters deemed necessary for waste characterization. Unit samples are analyzed for only those constituents requiring management for proper disposal.

6

PROTOCOL FOR CHEMICAL ANALYSES

SOLUBLE CONSTITUENTS ANALYSES

Pure water is not a good conductor of electricity; however, water containing soluble salts will conduct electricity roughly in proportion to the quantity of salt present. This relationship is accurate enough to determine the salt concentration by measuring the electrical resistance or the electrical conductivity of a solution. In this procedure, the electrical conductivity of a soil extract is used to estimate the amount of soluble salts in a soil.

To determine the soluble salts, the soil is extracted with distilled or deionized water. The salts dissolved will increase with increasing soil:water ratio. Salt determinations normally are based on conductivity of an extract of a soil when at saturation, called saturated paste extract.

Measurements of soluble salts by this method are accurate and reproducible in the range usually occurring in the saturation extracts. There is a high degree of positive correlation between conductivity, total concentration of cations or anions, and the osmotic pressure of soil-water extracts. However, due to differences in equivalent weights and equivalent conductivities of the different ions, the relationship of conductivity to salt content on a weight-to-weight basis is poor. The following electrical conductivity relationships are frequently employed:

- Salt concentration, mg/l = 613 × electrical conductivity, millimhos/cm *(the constant 613 produces very good correlation for oil and gas field wastes)*
- Total cation concentration, meq/l = 10 × electrical conductivity, millimhos/cm
- Osmotic pressure, atm = 0.36 × electrical conductivity, millimhos/cm

The standard unit for conductivity, mhos/cm is too small for field use. Accordingly, for convenience millimhos/cm are used. Units commonly reported are:

0.001 mhos/cm = 1 millimhos/cm = 1,000 micromhos/cm

Preparation of Saturated Paste and EC Analysis

Place 100g of soil into a 250ml beaker. Add water to the soil in small increments from a graduated cylinder. Successive increments of water are added *without vigorous stirring of the soil* until water just wets the entire mass, then a few drips more are added slowly until the surface glistens slightly. After this moisture content has been reached, the soil is stirred with a glass rod and drops of water are added until the soil is a "thin paste" which just barely flows together to close around a hole left by the rod. The soil is now at the moisture saturation percentage. The surface of the water saturated soil glistens, and the air has been excluded. *Record the amount of water used to reach this point* Calculate the percent of moisture at saturation:

$$\text{\% Moisture at Saturation} = \frac{\textit{Wt. Soil Water}}{\textit{Wt. oven dry soil}} \times 100$$

Allow the soil to stand for about 30 minutes and transfer to a Buchner funnel provided with filter paper. Apply suction and collect at least 10 ml of extract. Take the temperature of the extract and set the proper correction on the conductivity meter. Transfer the extract to a 50 ml beaker. Rinse the meter cell with distilled water several times and with the soil extract once before taking a conductivity reading. Obtain a reading on the solu-bridge. Report the conductivity of the soil in millimhos/cm (mmhos/cm). Further details on obtaining the saturated paste extract and EC analysis are given in Appendix I, Annexes A and B, respectively.

Peripheral Analyses and Relationships

The "equipotential" moisture status associated with a saturated paste condition represents the highest likely moisture level in the field. It is a reproducible value but requires some training to perfect. Unskilled technicians tend to puddle the soil by adding too much water. Puddled soils imbibe water to varying degrees causing considerable deviation in replicate parameter analysis. Drilling fluids received by the laboratory for waste analyses often contain smectitic clays in a puddled condition. Puddled soils can have in excess of 200-300 moisture and represent a significant dilution of salts and gross underestimation of the salinity hazard associated with the waste. To avoid this problem, a subsample of drilling fluid should be air dried and then moistened to a saturated paste level or extracted at a 1:1 soil: water ratio (by weight) for soluble constituent analyses.

Some matrices may be hydrophobic (repel water) due to the presence of petroleum hydrocarbon, and require heat treatment for hydrocarbon removal prior to sample preparation for soluble constituent analyses. For details on testing for hydrophobicity and heat treatment see Appendix I, Annex C.

Soil Reaction

Soil reaction is an estimate of the hydrogen ion concentration as measured with a combination glass electrode and pH meter. The procedure for determining pH is given in Appendix III, Annex A. A saturated paste provides a thick moisture film and good contact with the glass electrode. The normal pH range for good growth of agronomic crops is 5.5 to 8.3. Generally, soils with a pH below 5.5 develop toxic aluminum and manganese levels and soils pH above 8.3 reflect a sodium bicarbonate/carbonate buffer system resulting in excessive and plant toxic soluble sodium levels. Many native plants have adapted to soils with a pH range of 4 to 9. Nevertheless, we recommend working to a pH range of 5.5 to 8.3. Excessive soil reactions are readily treated by amendment Chapter VIII.

Excessive alkalinity (pH>8.3) typically accompanies sodic soil conditions. This condition requires sulfur amendment to lower the soil reaction, and gypsum to reduce sodium saturation in the adsorbed phase. Excessive acidity (pH <5.5) is controlled with lime. Procedures for determining the alkalinity and acidity of E&P wastes are given in Appendix III, Annexes B and C, respectively.

Soluble Cations

Sodium (Na), potassium(K), calcium(Ca), and magnesium (Mg), are the cations of interest. They are measured for saturated paste extracts using direct aspiration atomic absorption spectroscopy. Sodium and potassium are analyzed in a flame emission mode. Calcium and magnesium are analyzed in the absorbency mode. Results are normally reported in milliequivalents per liter (meq/l).

Sodium adsorption ratio (SAR) is calculated from saturated paste concentrations expressed in meq/liter:

$$SAR = Na\ /[(Ca + Mg)/2]^{1/2}$$

SAR is used in conjunction with EC to evaluate the potential hazards associated with sodium and soluble salts. A soil with an SAR value greater than 15 tends toward a sodic soil condition. A sodic soil is defined as one with greater than 12 % of the exchange sites occupied by sodium and reflects an undesirable physical status (a structureless or dispersed soil condition). The degree of hazard to plants is related to the ionic strength of the matrix. Electrical conductivities (EC) greater than 4 mmhos/cm are classified as saline. Soils with a high SAR (15 % and low salinity level (EC <1) are readily ameliorated with gypsum amendments. When both SAR and EC values are high (> 15 and 4, respectively), the soil is considered saline-sodic and requires both gypsum and leaching to restore productivity. The procedure for determining soluble cation distributions from which SAR is calculated is given in Appendix I, Annex D.

Soluble Anions

Bicarbonates, carbonates, hydroxides, chlorides, and sulfates are the anions of interest in soils, waste drilling fluids, and leachates. Bicarbonate carbonate, and hydroxide ions are titrated in the presence of phenolphthalein and methyl orange indicators when color changes are readily visible in the various water extracts. Water extracts from sodic soils and some drilling fluids are often darkly colored and require titrations by a pH meter instead of by indicators. Titration of an alkaline solution to the phenolphthalein end point (pH 8.2) with a standard solution of strong acid provides a measure of carbonate, and further titration to the methyl orange end point (pH 4.5) provides a measure of bicarbonate. Chloride is measured for noncolorgenic soil extracts and water samples with an argentometric titration using potassium chromate as the indicator. The potentiometric method is used for colored or turbid extracts. Drilling fluids containing chromates and organics may require pretreatment by "soft digestion" to remove interferences prior to titration. Sulfate is determined for soil and drilling fluid extracts using the turbidometric method. Background correction for color or turbidity is made by subtracting the absorbency values in the absence of barium chloride.

Quantitative and Quality Control

Quality assurance and the confidence in the reliability of analytical results for soluble constituents begin with deionized water in standard formulations and dilutions, and as a final rinse of all laboratory glassware. Scrupulous cleaning of all glassware starts with a pre-rinse scrubbing with soap and water, rinsing with distilled water, dilute nitric acid, and finally deionized water.

Analytical and electronic balances must be calibrated and checked against standard weights prior to and during use in gravimetric assays. Calibrated glassware is used in volumetric procedures.

Precision of direct reading instruments is maintained by routine calibration against commercially available buffer solutions and standards. Cations and anions are determined by indirect methods employing a comparison of the sample matrix response to a standard curve prepared by known addition technique. Precision and accuracy are evaluated statistically with replicate determinations and spike recovery analyses. Detection and sensitivity limits are evaluated with reagent blanks and standard deviation about the mean for the low standard. Matrix interferences for alkali metal determinations are controlled by the addition of suppressants to aid in element ionization and surfactants that maintain optimum atomization rates into the instrument flame.

SOLID PHASE ANALYSES

Colloidal clay and organic particles combine to form the chemically active components of soils and E&P waste solids. Colloidal size fractions have very large surface areas. Additionally, colloids have an electrical charge (CEC) emanating from ionic substitution and broken crystal edges. Soil clays are comprised of crystalline sheets of silicon oxides and aluminum oxides or hydroxides bound to each other by shared oxygen atoms (Chapter 2). To reiterate, a charge deficiency within the crystalline lattice arises when aluminum (Al^{+3}) replaces silicon (Si^{+4}) in a tetrahedral sheet, and iron (Fe^{+2}), zinc (Zn^{+2}), and magnesium (Mg^{+2}) replace aluminum (Al^{+3}) in an octahedral sheet. The negative charge deficiency attracts and retains cations. Cations are free to exchange between soil solution and the adsorbed phase. The negative charge per unit soil mass is termed the cation exchange capacity (CEC). An equilibrium exists between soluble and adsorbed ions since cation exchange is rapid and reversible. The equilibrium relative to the solid phase is termed the exchangeable cation distribution.

Cation Exchange Capacity (CEC)

Soil colloids have both negative and positive charged surfaces. Capacity to adsorb cations per unit mass is referred to as the cation exchange capacity (CEC), and is expressed in milliequivalents per 100 g soil mass on an oven dry (105°C) weight basis. The capacity to adsorb anions is called the anion exchange capacity (AEC) and is expressed in milliequivalents per 100 g soil mass.

The two most common measures of CEC are the quantity of cations adsorbed from a salt solution buffered at either pH 7 or 8. CEC determinations from salt solutions buffered at pH 8 give an accurate assessment of the charge at the clay surface and a reasonable measure of contributions from organic matter and non-crystalline colloids. Most E&P wastes are alkaline in reaction and therefore best evaluated using pH 8 buffered salts.

Exchangeable Cation Distribution

The exchangeable cations in a soil are the positively charged ions held on the surface exchange sites and in equilibrium with the soil solution. The major cations Na, Ca, K and Mg are called basic cations and the % of the CEC occupied by these cations is called the percent base saturation. Fertile soils approach 80% base saturation with bulk saturation distributed mainly as Ca and Mg.

Problems arise when the exchangeable Na saturation percentage (Na/CEC)*100 exceeds 12-15% of the CEC. These are termed sodic soils and are characteristically structureless and impervious to air and water. Many drilling fluids are dispersed with sodium. A major concern in site reclamation and disposal of waste drilling fluids is management of adsorbed sodium.

Procedures

The procedures for determining CEC and exchangeable cation distributions are given in Appendix I, Annexes E and F, respectively.

Quality Control and Assurance

Accuracy and precision of CEC and exchangeable cation distribution measurements are evaluated at a procedural level since they are influenced by treatment parameters, including incorporation of waste streams in natural soil bodies.

Accuracy assessment of CEC results from multiple determinations of laboratory reference materials representative of the matrix in question, and previously evaluated by several techniques. Precision is achieved by replicate determinations.

The 3 steps of CEC determination are:

- saturation of cation exchange sites with a specific cation;
- removal of excess saturating solution; and
- replacement of saturating cation.

Errors can arise at each of the three steps. Step 1 errors are minimized through selection of appropriate saturating cation and adequate rinsings with index cation solution. Many drilling fluid wastes contain adsorbed sodium initially rendering it the most suitable index cation. Step 2 errors are minimized by thoroughly washing of the sample with deionized water followed by isopropyl alcohol. Loss of sediment on decantation experienced with a total deionized water washing is reduced by using isopropyl alcohol in the final rinse process. Errors can arise during step 3 from samples from relatively unweathered soils due to carbonates or gypsum when calcium or magnesium are used as index and replacing cations. A sodium index and nitrate replacing agent negates possible error in step 3 caused by carbonates or gypsum.

TOTAL METALS ANALYSES

Most soils and E&P wastes contain low quantities of heavy metals in water soluble or exchangeable form. Accordingly, water or other extractants that partition (separate from the matrix) these fractions do not provide a suitable index for metals management. On the other hand, total metal analysis provides usable numbers and a basis for comparison with acceptable total metal loading criteria. The reader is referred to API Publication Number 4600 for technical guidance associated with land-managed E&P wastes.

Acid Digestion

The recommended protocol for total metal digestion is detailed in Appendix II. All total metals concentrations, except barium, are estimated accurately by metals fractionation procedure. A separate procedure was developed for True Total Barium (Appendix II).

Quality Control and Assurance

Matrix variability is managed by careful blending prior to weighing subsamples for analysis. Certified reference soils and sediments are included along with duplicate digests and matrix spikes to evaluate precision and accuracy of analysis.

ORGANIC ANALYSES

The oil and grease (O&G) procedure (Appendix I, Annex H) and TPH-IR (EPA 418.1 as modified for soils and solid wastes) provide an index of petroleum hydrocarbon partitioned from the matrix by a solvent. The

O&G method uses methylene chloride and the TPH-IR procedure uses freon. Sample matrices are acidified to reduce co-extraction of naturally occurring organic matter, and in the TPH-IR method passed through silica gel to remove interferences.

Either of these analyses provides a reasonable estimate of total petroleum hydrocarbon. Values are used to determine hydrocarbon loading rates with respect to on-site land treatment or bioremediation, and also serve as an index to other management imputs. Specific component analyses may be necessary for permit purposes, depending on E&P waste source and regulatory agency involved.

Neither O&G nor TPH serve as an adequate index for tracking bioremediation treatment performance. Both methods give inflated petroleum hydrocarbon values. We believe the reason for this is the conversion of petroleum hydrocarbon substrates to co-extracted biomass. Gas chromatographic analysis of total petroleum hydrocarbon standardized to peak area response is a better indicator of bioremediation performance. (Deuel and Holliday, 1993)

Quality Control and Assurance

Duplicate analysis, spike recovery analysis, reagent blanks, and methods blanks are necessary to evaluate precision and accuracy of the analyses performed. Selection of standards of close similarity to the source hydrocarbon is critical to accurate hydrocarbon analyses. This is particularly true when comparing peak area response in gas chromatography.

References

Deuel, Lloyd E., Jr. and G.H. Holliday. 1993. "Determining Total Petroleum Hydrocarbons in Soil". Presented at the 68th Annual Technical Conference and Exhibition of the Society of Petroleum Engineers SPE 26394, held in Houston, Texas, October 3-6.

7

LIMITING CONSTITUENT CRITERIA

INTRODUCTION

A review by the U.S. Environmental Protection Agency (EPA, 1987) of oil and gas production activities and analysis of wastes generated, concluded reserve pit solids are not hazardous in the sense of being dangerous or toxic. Their conclusions confirm the Industry's own research efforts and study results.

By comparing regulations with what is known about E&P wastes, EPA concluded E&P activities have little adverse impact on the environment. However, only one state, Louisiana, mandates treatment protocol and demonstration of treatment effectiveness by comparing post closure analyses to regulatory threshold values.

The vast majority of wastes generated by exploration and production are not hazardous. However, most require some form of management to preclude adverse impact to land resources, land use, or water resources.

Salts and the dispersed nature of waste solids (lack of physical structure) are the principal limitations associated with oil field wastes. Some wastes contain elevated hydrocarbon levels, while others are high in salts. State regulations require pit management and closure in a timely manner combined together to minimize the potential for an adverse impact.

STATE REGULATIONS

Most states either exempt reserve pit construction from regulations or provide demonstration options to operators allowing unlined earthen pit use during drilling. Demonstration may simply take the form of disclosing the anticipated mud program.

Standard operating procedure for pit closure in most states includes dewatering, compacting, and backfilling. Some specify the thickness of the final cover and material to be used, usually stripped and segregated topsoil. Liners are required in environmentally (hydrologically) sensitive areas when other than fresh water muds are used. Some states do not require liners but substitute more stringent timing for liquid removal and closure. All states require closure following cessation of drilling operations unless the pit has been designed and constructed for multiple use. A summary of selected state guidelines or regulations for pit closure is given below:

Colorado

A. **Regulatory Authority/Jurisdiction**
   1. Department of Natural Resources
      a. Oil and Gas Conservation Commission
      b. Exploration & Production Waste Management

B. **Pit Construction/Management**
   1. Earthen pits allowed
   2. No permit is required for fresh water base mud reserve pits
   3. Permits are required for reserve pits which will contain salt-base or oilbase drilling fluids or where the well encounters a salt section which would increase salt content of the drilling fluid to >50,000 mg/l TDS
   4. Pit application must be submitted with a filing and service fee of $200 and must include:
      a. Drilling location layout plan
      b. Pit size
      c. Type of mud program
      d. Anticipate use time
      e. Mud disposal plan
      f. Topographic map (1:24,000) indicating drainage and distance to bodies of water
      g. Subsurface type
      h. Type of pit sealing materials including specification and method of application
   5. Earthen pits prohibited on fill
   6. Top soil removed in the construction of the pit must be stockpiled and used for pit reclamation of the pit site
   7. Less than 3 ft of freeboard is prohibited
   8. Soil liners must be at least 6 in thick after compaction, cover pit bottom and sides and have coefficient of hydraulic conductivity of at least 1.0×10^{-7} cm/sec

9. Geomembrane liners must have a minimum of 12 mils
10. Remove fluids and close pit within 6 months unless director allows 6 additional months

C. Acceptable Pit Closure

1. On-site disposal
   a. Land owner agreement required
   b. Waste streams to be land treatment must be analyzed for:
      1) Benzene, toluene, ethylbenzene and xylenes (BTEX)
      2) TPH
      3) Oil and grease
      4) TCLP for heavy metals
      5) TDS
   c. Oil and grease concentration of waste may not exceed 50,000 mg/kg
   d. Walk-away concentrations after waste incorporation (Table 7-1)

Table 7-1 Colorado threshold remediation levels

Parameters	Regulatory Threshold
pH	6.5-9.0 s.u.
oil and grease	10,000 mg/kg
arsenic	100 mg/kg
barium	2,000 mg/kg
cadmium	20 mg/kg
chromium	100 mg/kg
mercury	4 mg/kg
selenium	20 mg/kg
silver	100 mg/kg

2. Off-site disposal
   a. Waste streams must be analyzed for:
      1) Benzene, toluene, ethylbenzene and xylenes (BTEX)
      2) TPH
      3) Oil and grease
      4) TCLP for heavy metals
      5) TPH

Louisiana

A. **Regulatory Authority/Jurisdiction**

1. Department of Natural Resources
    a. Office of Conservation - Injection and Mining Division
    b. Department of Environmental Quality
    c. Coastal Management Section

B. **Pit Construction/Management**

1. Earthen pits allowed
2. Reserve pit notification satisfied by application for drilling or workover permit
3. Commingling produced waters or oil with reserve pit wastes prohibited
4. Protect pits from surface waters by levees, walls and/or drainage ditches
5. Less than two feet of freeboard prohibited
6. Empty fluids and close pit within six months after cessation of drilling

C. **On-site Closure Requirements**

1. Requires land owner's permission
2. Acceptable disposals include:
    a. Land treatment
    b. Burial or trenching
    c. Solidification
3. Annular injection allowed by permit only
4. Adequate treatment demonstrated by analysing closed pit soil mixture according to prescribed analytical protocols and comparing results with acceptance threshold criteria (Tables 7-2 and 7-3)
5. Reserve pits for wells (5,000 ft deep are exempted pre- and post-closure testing, if only fresh water mud contains (25 lb/bbl bentonite, (0.5 lb/bbl caustic soda, and (50 lb/bbl barite

- Top of buried mixture must be (5 ft below ground level and covered with at least 5 ft of background soil. Also, the bottom of the buried mixture must be at least 5 ft. above the seasonal high water table.
- Louisiana also permits reserve pit waste disposal by solidification, passive closure, and off-site disposal.

Table 7-2 Louisiana threshold remediation levels for land treatment

Parameter	Regulatory Threshold
pH	6.0-9.0
Electrical Conductivity (EC)	<4.0 mmhos/cm(uplands) <8.0 mmhos/cm (elevated wetlands)
Sodium Adsorption Ratio (SAR)	<12.0
Exchangeable Sodium Percentage (ESP)	<15.0%
Oil & Grease	<1.0%
Total Metals (mg/kg)	
Arsenic	10
Barium (Wetlands)	20,000
Barium (Uplands)	40,000
Cadmium	10
Chromium	500
Lead	500
Mercury	10
Selenium	10
Silver	200
Zinc	500

Table 7-3 Louisiana threshold remediation levels for dilution burial

pH and Metals	As above
Electrical Conductivity (EC)	<12.0 mmhos/cm
Moisture Content	<50% by weight
Sodium Adsorption Ratio (SAR)	NA
Exchangeable Sodium Percentage (ESP)	NA
Oil & Grease	3.0%

New Mexico

A. Regulatory Authority/Jurisdiction

1. New Mexico Oil Conservation Division

B. Pit Construction/Management

1. Earthen pits allowed
2. Closure plan required and approved by state prior to commencing closure operations
3. Site assessment
   a. Depth to groundwater

b. Horizontal distance from site to springs, wells used by less than 5 households for domestic and stock purposes
c. Horizontal to surface water bodies; i.e., rivers, streams, creeks, irrigation canals and ditches, lakes, and ponds
d. Native soil type
   1) low permeability
   2) moderate permeability
   3) high permeability
e. Degree of contamination

Highly contaminated/saturated soils

1) Contain a free phase hydrocarbon liquid or exhibit gross staining or emit very strong hydrocarbon odor (none of these terms are defined).

Unsaturated contaminated soils

1) Contain no free hydrocarbon liquid but contain benzene, toluene, ethylbenzene and xylenes and TPH above background levels. We interpret this to mean BTEX and TPH both must be above background concentrations.
   f. Groundwater quality
4. TPH contamination determination methods
   a. E&P exempted waste impoundments
      1) Filter Paper Method - Place sample of contaminated soil on laboratory filter paper. If liquid hydrocarbons pass through the filter paper as an immiscible phase, the soil is a Highly Contaminated/Saturated Soil
      2) Physical Observations - Study a representative sample of contaminated soil for:
         - free petroleum hydrocarbons
         - gross staining or
         - strong hydrocarbon odors

      If any one of these characteristics is exhibited by the soil, the soil is considered highly contaminated/saturated soil

      3) Headspace Analysis - Sniff the headspace gas of a half full liter container of contaminated soil at 59(- 77(using a Photo Ionization Detector (PID) or Flame Ionization Detector (FID) calibrate for benzene response *(PID and FID produce different readings for the same soil. Additionally, the same hydrocarbon concentration produces differing instrument responses depending on soil type).* Record the response. New Mexico guidelines provide no standard of acceptance against which to compare the PID or FID instrument responses.

4) Analytical Methods - Analyze for:
   - BTEX using EPA Method 602/8020
   - TPH using EPA Method 418.1

5. Ranking criteria - sum the individual scores for the 4 characteristics

	Score
a. *Depth to groundwater*	
<50 ft or unknown	20
50 - <100 ft	10
100 - 200 ft	5
>200 ft	0
b. *Wellhead protection area*	
<1,000 ft from a water source or <200 ft from a private domestic water source?	
Yes	20
No	0
c. *Distance to surface water*	
<500 ft	20
500 - 1,000 ft	10
>1000 ft	0
d. *Native soil type*	
Low permeability	0
Moderate permeability	5
High permeability	10

6. Soil remediation levels
   a. Highly contaminated/saturated soils
      1) Excavate until representative sample of walls and bottom is less than "Total Ranking Score," Table 7-4

Table 7-4 New Mexico threshold remediation levels

Total Ranking Score	**<10**	**10-19**	**>19**
Contaminate		**Maximum Allowable Concentration (mg/kg)**	
Benzene*	10	10	10
BTEX*	50	50	50
TPH**	5,000	1,000	10

* Benzene, toluene, ethylbenzene and xylenes EPA Method 602/8020

** TPH EPA Method 418.1

Author's Note: TPH; levels by IR (Method 418.1) probably are unattainable for weathered hydrocarbons because IR records humus as TPH.

- *excavating to maximum depth and horizontal extent practicable, i.e. maximum reach of a backhoe*
- *insitu soil treatment, i.e., vapor renting, bioremediation or other treatment: no threshold level specified*

b. Unsaturated contaminated soil
   - remediated to appropriate Total Ranking Score, Table 7-4

7. Remediation methods
   a. Landfarming - onetime application of contaminated soil may be land treated if the site Total Ranking Score is (19. The contaminated soil must be spread no deeper than 6 in. per lift within a bermed area. The application must be disked at least once per month. Moisture and nutrients may be added. For land treatment site receiving a Total Ranking Score >19 must use an impervious liner within the bermed area.
   b. *Insitu* soil treatment - vapor venting or bioremediation
   c. Alternate methods - other methods acceptable to the Oil Conservation Division

North Dakota

A. Regulatory Authority/Jurisdiction

1. Industrial Commission of North Dakota/Oil & Gas Regulatory Division
2. U.S. Forest Service
3. Operates under general rules and regulations

B. Pit Construction/Management

1. Earthen pits allowed
2. Liners usually required—geology and geohydrological setting:
   a. Permeability of materials encountered
   b. Proximity to surface and groundwater resources
   c. Anticipated mud program
3. Conservation of topsoil
   a. Removal of top 8 inches of soil
   b. Stockpile
   c. Redistribute as final ground cover
4. On-site disposal of waste solids
   a. Notice of closure with state geologist (Form 4)
   b. Remove liquids

        1) Another lined pit
        2) Injection well
        3) Sprayed on roads for dust control
    c. Trench burial or backfilled
    d. Redistribute topsoil
    e. Revegetation required

C. Time and Notification Requirement

1. Pits are usually closed within one year following cessation of drilling
2. State geologist monitors restoration actively

Oklahoma

A. Regulatory Authority/Jurisdiction

1. Oklahoma Corporation Commission
   Pollution Abatement Department
   Oil & Gas Conservation Division
2. Regulated under Oklahoma Rules 165:10-1-2, 165:10-7-19 and Appendix I

B. Pit Construction/Management

1. Earthen pits allowed
2. Soil or geomembrane liner requirements
    a. Geology and geohydrological considerations
        1) Depth to groundwater within 6 ft of pit bottom
        2) Liquid chlorides anticipated to exceed 5,000 mg/liter
        3) Oil base mud
        4) Multiple use pit
    b. Liner must be chemically compatible with pit contents and possess ultraviolet light protection
    c. Soil liner permeability less than 1x10-6 cm/sec
    d. Fluid levels maintained 2 ft below berm top
3. On-site one time disposal of waste solids is allowed
    a. Dewater pit and backfill
    b. Solidification
    c. Annular injection (requires permit)
    d. Land application (requires permit)
    e. Trenching, except in hydrologically sensitive area or oil base pits
4. Off-site disposal to permitted commercial facilities allowed

C. **Time and Notification Requirement**
1. Earthen pits closed within 6 months after drilling completed
2. Lined pits closed within 12 months after use ceases
3. Notification required to field inspector or District Office
   a. 48 hours prior to closure
   b. 48 hours after reclamation

D. **One Time Land Application of Water-base Fluids from Earthen Pits Allowed**
1. Site suitability
   a. Land slope
      1) 8 % maximum for spray irrigation method
      2) 5 % maximum for other application methods
   b. Soil depth; 20 in or more to bedrock
   c. Soil texture; at least 12 in cumulative or individually of loam, silt loam, silt, sandy clay loam, silty clay loam, clay loam, sandy lay, silty clay, or clay
   d. Soil salinity
      1) (4 mmhos/cm in upper 6 in of soil
      2) ESP < 10%
      3) TDS = EC (640
   e. Depth to water table
      1) >6 ft from soil surface
   f. Distance from water bodies
      1) (100 ft from site boundary to perennial stream, and
      2) (50 ft from site boundary to intermittent stream shown on USGS 15-second map or freshwater pond, lake or wetland as shown on Fish and Wildlife Wetlands Inventory Map
2. Sampling requirements
   a. Notice to district - Notify field inspector at least 2 days before sampling
   b. Receiving soil - 4 samples (0-6 in deep) and (12-20 in deep) per 10 acres, or part thereof, composited into 2 pint samples by depth
   c. Drilling fluids/cuttings
      1) No freshwater, except natural precipitation, shall be added to reserve pit
      2) 5 samples per 25,000 bbls or less, plus
         1 sample per 5,000 bbls over 25,000 bbls
      3) Composite above samples into 2 qt foil or teflon covered glass containers

4) Deliver composite samples to laboratory within 7 days

3. Sample analyses
   a. Samples shall be analyzed by a laboratory certified by the Oklahoma Water Resources Board using either 1:1 (by weight) extract or saturated paste extract
   b. Receiving soil analysis
      1) Electrical Conductivity (EC)
      2) Exchangeable Sodium Percentage (ESP)
   c. Drilling fluid/cuttings analyses
      1) Electrical Conductivity (EC)
      2) Oil and Grease (O&G)
      3) Solids dry weight
4. Permit application
   a. Only the well operator may apply for a land application permit
   b. Application form
      1) Submit Form 1014S and three copies to the Field Inspections Office and
      2) Attach one copy each of
         a) Surface owner's written permission to land apply pit waste; use Form 1014 L
         b) Scale diagram of application area
         c) Site suitability report
         d) Analysis of drilling fluid/cuttings
         e) Analysis of soil samples, and
         f) Loading calculations
5. Loading calculations. Refer to loading worksheet below.
6. Land application method
   a. Apply about 4 in thick
   b. If more than 500 lb/ac of oil and grease and/or 50,000 lb of solids are applied, disk and cross disk to 6 in deep

TOTAL DISSOLVED SOLIDS (TDS)

EC of receiving soil ______ mmhos/cm x 640* = ______ mg/l TDS
TDS in receiving soil ______ mg/l x 2 = ______ lbs/ac TDS in receiving soil
6,000 lbs/ac TDS* - _______ lbs/ac TDS in receiving soil = *Maximum TDS* (lbs/ac) to be applied ______

EC of materials to be applied ______ mmhos/cm x 640* = ______ mg/l TDS
Maximum TDS (lbs/ac) to be applied ______ ÷ (TDS of materials to be applied

______ mg/l x 0.000001) = *Maximum weight* of materials to be applied ______lbs/ac

For liquid materials:
Maximum weight of materials to be applied ______ lbs/ac ÷ (sample weight ______ lbs/gal† x 42) = *Maximum loading* ______ bbls/ac
Total volume of materials to be applied ______ bbls ÷ *Maximum loading* ______
bbls/ac = *Minimum acres* required ______

For solid materials:
Maximum weight of materials to be applied ______ lb/ac ÷ (sample weight ______ lbs/gal† x 202) = *Maximum loading* ______ cu yds/ac
Total volume of materials to be applied ______ cu yds ÷ *Maximum loading* ______ cu yds/ac = *Minimum acres* required ______

CHLORIDES (Cl)

Cl in receiving soil ______ mg/kg x 2 = ______ lbs/ac Cl in receiving soil
3,500 lbs/ac Cl** - _______ lbs/ac Cl in receiving soil = *Maximum Cl* (lbs/ac) to be applied ______
Maximum Cl (lbs/ac) to be applied ______ ÷ (Cl of materials to be applied ______ mg/kg x 0.000001) = *Maximum weight* of materials to be applied ______ lbs/ac

For liquid materials:
Maximum weight of materials to be applied ______ lbs/ac ÷ (sample weight ______ lbs/gal† x 42) = *Maximum loading* ______ bbls/ac
Total volume of materials to be applied ______ bbls ÷ *Maximum loading* ______ bbls/ac = *Minimum acres* required ______

For solid materials:
Maximum weight of materials to be applied ______ lbs/ac ÷ (sample weight ______ lbs/gal† x 202) = *Maximum loading* ______ cu yds/ac
Total volume of materials to be applied ______ cu yds ÷ *Maximum loading* ______ cu yds/ac = *Minimum acres* required ______

OIL AND GREASE (O&G)

40,000 lbs/ac O&G ÷ (O&G of materials to be applied ______ mg/l x 0.000001) = *Maximum weight* of materials to be applied ______ lbs/ac

For liquid materials:
Maximum weight of materials to be applied ______ lbs/ac ÷ (sample weight ______ lbs/gal† x 42 gal/bbl) = *Maximum loading* ______ bbls/ac
Total volume of materials to be applied ______ bbls ÷ *Maximum loading* ______ bbls/ac = *Minimum acres* required ______

For solid materials:
Maximum weight of materials to be applied ______ lb/ac ÷ (sample weight ______ lbs/gal† x 202) = *Maximum loading* ______ cu yds/ac
Total volume of materials to be applied ______ cu yds ÷ *Maximum loading* ______ cu yds/ac = *Minimum acres* required ______

DRY WEIGHT

Wet weight of drilling mud ______ lbs/gal x ______ % dry solids = ______ lbs/gal dry weight†
______ lbs/gal dry weight x 202 = ______ lbs/cu yd
200,000 lbs/ac Dry Weight ÷ ______ lbs/cu yd = *Maximum cu* yds/ac ______
Total volume of materials to be applied ______ cu yds ÷ *Maximum cu* yds/ac ______ = *Minimum acres* required ______

* *640 required by state.*
** *Soil loading standards are based upon standards set forth in "Diagnosis and Improvement of Saline and Alkaline Soils". U.S. Agriculture Handbook 60. 1954. U.S. Printing Office, Washington, D.C.; Moseley, H.R., 1983. "Summary and Analysis of API Onshore Drilling Mud and Produced Water Environmental Studies", American Petroleum Institute Bulletin D-19.*
† *Based on dry weight percent of composite sample of materials. (Source: Added at 12 OK Reg. 2039, 7-1-95).*

Texas: Pits

A. Regulatory Authority/Jurisdiction
1. Texas Railroad Commission - Oil & Gas Division
2. Regulated under Statewide Rule 8, Water Protection

B. Pit Construction/Management
1. Earthen pits allowed
2. Reserve pits and mud circulation pits allowed without a permit

   a. Drilling fluids allowed
      1) Fresh water base
      2) Salt water base
      3) Oil base
   b. Drill cuttings, sands and silts
   c. Drill pipe and rig wash water
   d. Drill stem test fluids
   e. Blowout prevention test fluids
3. Closure requirements
   a. Fresh water muds
      1) Chloride <3,000 mg/l in liquids
      2) On-site or off-site land treatment or disposal
      3) Land treatment with land owner's permission
      4) Dewater, backfill, and compact
      5) Complete closure within one year following cessation of drilling
   b. Salt water muds
      1) Chlorides (3,000 but (6,100 mg/l)
      2) On-site land treatment only
      3) Land treatment with land owner's permission
      4) Complete closure within one year following cessation of drilling
   c. Salt water and oil base muds in liquids
      1) Chloride >6,100 mg/l
      2) On-site land treatment
      3) Land treatment with land owner's permission
      4) Dewatered within 30 days
      5) Backfill and compact within one year following cessation of drilling

Pit closure represents voluntary compliance with landowner serving as watch dog. During trials where this book is entered as an exhibit, lawyers have stated that nowhere is there any control over pit closure in Texas (GHH).

Texas: Soils

A. Regulatory Authority

1. Texas Natural Resources Code Section 91.101
2. Regulated under Statewide Rule 91

B. Management

1. Remediation of crude oil contaminated soil
2. Applies to soils contaminated after effective date of Rule 91

3. Effective date 1 January 1994
4. Remediate on-site or remove to off-site disposal
5. Action level, immediately
   a. Mix in place to 5 % by weight TPH or
   b. Remove to off-site disposal site
   c. Remove to secure interim storage for future remediation or disposal
6. Final cleanup level, 1% TPH achieved within one year

C. **Notification**
1. Report immediately if >5 bbls released to soil (Form H-8)
   a. Area of contamination
   b. Depth of contamination
   c. Volume of soil contamination >1% TPH
2. Signed statement indicating soil contamination >1% TPH was remediated or disposed; paper towel test allowed
3. Signed statement indicating soil contamination >5% TPH was mixed, disposed, or placed into interim storage
4. Detailed description of the disposal or remediation method used or planned
5. Estimated date of completion of site cleanup; if release is >10 bbls, provide final analysis of samples verifying cleanup level achieved
6. Spill of 5 bbls or less must be remediated to 1% or less within one year; no report is required
7. Spill >25 bbls must be reported(Form H-8) as above and the operator must submit a final report upon completion of site remediation

Utah (1994 Guidelines)

A. Regulatory Authority/Jurisdiction: Division of Oil, Gas and Mining

B. Pit Construction: Earthen pits allowed

C. **Ranking System**
1. Applies only to hydrocarbon contamination; no guidelines for salinity
2. Cleanup level depends on total score
3. Distance to groundwater from surface

Distance, feet	*Score*
>100	0
100 to 75	4
<75 to 50	8
<50 to 25	12
<25 to 10	16
<10 and recharge area	20

4. Soil type

Type	*Score*
Low permeability (clay, shale, fat clay etc) as specified	0
Moderate permeability *(clayey sand,* poorly graded sand-clay mixtures, etc.) as specified	10
High permeability *(fine sand, silty sand, sand, etc.)* as specified	20

5. Precipitation

Quantity (inches/year)	*Score*
<10	0
10 to 20	5
>20	10

6. Distance to municipal water well

Distance (feet)	*Score*
>5,280	0
5,280 to 1,320	6
<1,320	8
<500	10

7. Distance to other water well

Distance (feet)	*Score*
>1,320	0
<1,320 to 300	5
<300	10

8. Distance to surface water

Distance (feet)	*Score*
>1,000	0
1,000 to 300	5
<300	10

9. Population within 3 miles

Number of people	*Score*
<100	0
100 to 3,000	5
>3,000	10

10. Utility conduits within 300 feet

Presence	*Score*
not present	0
unknown	5
present	10

11. Score Summation

$$R=\sum_{S\ 1}^{8} R_S$$

12. Cleanup Level

Ranking	*Cleanup Level*
<50	Level I
(50	Level II

13. Cleanup Criteria

LevelAllowable TPH-IR above background level	
I	10,000 mg/kg
II	50/% benzene in TPH analysis

For example, if the TPH analysis indicates 0.01% benzene in waste TPH = 50/0.01 = 5,000 mg/kg allowable

14. No closure time specified

Wyoming

A. Regulatory Authority/Jurisdiction

1. Oil & Gas Conservation Commission
2. Bureau of Land Management
3. No specific rule or regulation

B. Pit Construction/Management

1. Earthen pits
2. Liners may be required due to
   a. Geology and geohydrological setting
   b. Depth of well
   c. Anticipated mud program
3. On site disposal of waste generated
   a. Liquid fraction evaporated
   b. Redistribute solids
   c. Backfill with berm material
   d. Replace topsoil and restore contour

C. Pits are usually closed within one year following cessation of drilling.

AMERICAN PETROLEUM INSTITUTE METALS CRITERIA

The American Petroleum Institute (API) published in 1995, (API, 1995) metals criteria for E&P wastes, (Table 7-5).

Table 7-5 API recommended E&P waste metals criteria

Metal	Extraction Method	Maximum Soil Concentration (mg/kg)
Arsenic	EPA Method 3050*	41
Barium	LDNR True Total Barium**	180,000
Boron	Hot Water Soluble (Carter, 1993)†	2 mg/L††
Cadmium	EPA Method 3050	26
Chromium	EPA Method 3050	1,500
Copper	EPA Method 3050	750
Lead	EPA Method 3050	300
Mercury	EPA Method 3050	17
Molybdenum	EPA Method 3050	see below
Nickel	EPA Method 3050	210
Selenium	EPA Method 3050	see below
Zinc	EPA Method 3050	1,400

* EPA, Testing Methods for Evaluation Solid Waste, SW-846, Third Edition

** Louisiana Department of Natural Resources, 1989. Laboratory Procedures for Analysis of Oilfield Waste, Statewide Order No. 29-B.

† Carter, 1993. Soil Sampling and Methods of Analysis, Boca Raton: Lewis Publishers, pp 91-93.

†† Guidance for boron is based on the soluble concentration with units of mg/L rather than the total concentration (mg/kg).

Molybdenum: On February 25, 1994, (59 FR 9050), EPA rescinded the risk-based maximum soil concentration for Mo of 9 mg/kg due to technical errors and established a non risk-based interim ceiling limit of 37 mg/kg. Under certain conditions this interim level may not be protective of grazing livestock. These conditions are alkaline soils under arid and semi-arid conditions with deficient levels of copper in the soil (see Discussion of Limiting Exposure Pathways).

Selenium: The limiting pathway concentration of 100 mg/kg was generated by EPA using the risk-based multipathway analysis (see Table 3).

However, the potential for plant uptake of Se may be high in alkaline soils under arid and semi-arid conditions. Plants that accumulate Se in these soils may pose a threat to grazing animals. Therefore, if elevated levels of Se are found in the waste, the operator should consider site conditions that control its availability.

PIT UTILIZATION AND WASTE MANAGEMENT (NOW DEFINED)

Non-hazardous Oilfield Waste (NOW) is defined as those liquids or solids generated in the exploration and production of oil and gas. A listing of such wastes is given below. These wastes were exempted by the 1980 amendment to the Resource Conservation and Recovery Act (RCRA) which regulates E&P wastes under Subtitle D. The U.S. Environmental Protection Agency (EPA) recommended the continuance of NOW exemption from Subtitle C of RCRA in the December 1987 report to Congress. Based on a recent EPA study, NOW most likely will not come under RCRA Subtitle C regulations.

Summary listing of non-hazardous oil field wastes

1. Produced brine or produced water
2. Oil base drilling mud and cuttings
3. Water base drilling mud and cuttings
4. Drilling, workover and completion fluids
5. Production pit sludges
6. Production storage tank sludges
7. Produced oily sands and solids
8. Produced formation fresh water
9. Rainwater from ring levees and impoundments at production and drilling facilities
10. Washout water generated from the cleaning of vessels (barges, tanks, etc) that transport non-hazardous oil field waste and are not contaminated by hazardous waste or material
11. Washout pit water from oilfield related carriers that are not permitted to haul hazardous waste or materials
12. Non-hazardous natural gas plant processing waste which is or may be commingled with produced formation water
13. Waste from approved salvage oil operators who only receive waste oil (BS&W) from oil and gas leases
14. Pipeline test water which does not meet discharge limitations established by the appropriate state agency, or pipeline pig water

15. Wastes from permitted commercial facilities
16. Material used in crude oil spill clean-up operations. State regulations are based on the assumption that oilfield wastes are non-hazardous and may be treated on-site with the landowner's permission. Treatment usually effects waste disposal and pit closure

SITE INDEXING

On-site waste disposal management is predicated on land resource impacted, waste constituent(s) requiring treatment, and intended land-use. Land resource designations include brackish wetlands, freshwater wetlands, and upland sites.

Further, upland sites are delineated according to intended land-use, including woodland/recreational and residential/farming distinctions. Depth to groundwater is an important parameter at upland sites where dilution burial is a disposal option. Wetland site indexing simply entails a land form designation followed by a descriptive modifier of the adjacent or encompassing water body (i.e., fresh or brackish/saline).

GUIDELINES FOR LIMITING CONSTITUENTS

Guidelines for limiting constituents (GLC) consist of criteria defining an acceptable range for NOW parameters: i.e.; salt, heavy metals and hydrocarbons. Those parameters exceeding limiting constituent levels generally require some level of treatment for on-site management. GLC values provide a guide for evaluating NOW materials. Typically, treatment or management of the most limiting characteristic (most excessive parameter) bring the other constituents below GLC values. GLC values for impoundments and remote field pits are presented in Table 7-6.

Pit Liquid Disposal

Drilling pits are used to store wash down water, storm water runoff, and drilling wastes. Pits normally store larger volumes of liquids than muds and cuttings. When drilling operations cease, the operator must dispose of these liquids prior to pit closure. Otherwise, liquids will be discharged to adjacent property during pit backfilling, and in some cases, damage land or enter water courses. In general, salinity and sodicity are not factors for closing pits in brackish or freshwater, submerged wetlands. The range given for EC and SAR in elevated wetlands reflects the range found under native conditions (Table 7-6). The upper limit suggested for pH is lower (pH 8 compared to pH 9) for land treatment under farming/residential land use than recreational land use. This reduces the potential for phosphorous and micronutrient deficiencies. The lower limits for pH of 6 and CEC of 15 are integrally related to the suggested total metal values. If a lower pH is required (e.g. pine forestation), one simply adjusts the total metals downward, or evaluates the potential impact of lowered pH on metals dissolution and release.

Table 7-6 GLC values utilizing land treatment and vertical dilution

		Wetlands		Uplands	
	Brackish	Freshwater		Forest/	Farming/
		Submerged	Elevated	Recreational	Residential
pH	6-9	6-9	6-9	6-8	6-8
EC (mmhos/cm)	–	–	8-50	4-8	2-4
SAR	–	–	14-100	<14	<12
CEC (meq/100g)	>15	>15	>15	>15	>15
ESP (%)	–	–	<25	<15	<15
Metals (mg/kg)		**Wetlands**		**Uplands**	
Arsenic		10		40	
Barium		20,000		40,000	
Cadmium		10		10	
Chromium		500		500	
Lead		500		500	
Mercury		10		10	
Selenium		10		10	
Silver		10		10	
Zinc		500		500	
Organics		**Wetlands**		**Uplands**	
Oil & Grease		1		2	

Petroleum hydrocarbon and salt are generally limiting constituents fordilution burial disposal (Table 7-7), and salt is the limiting factor for road spreading (Table 7-8).

Table 7-7 GLC values utilizing dilution burial for wetlands and uplands

Parameter	**Criteria**
pH	6-9
EC, mmhos/cm	<12
SAR	NA
CEC, meq/100g	NA
ESP, %	NA
Metals	**mg/kg**
Arsenic	40
Barium	40,000
Cadmium	10
Chromium	500
Lead	500
Mercury	10
Selenium	10
Silver	10
Zinc	500
Organics	
Oil & Grease	<3%

Table 7-8 GLC values utilizing road spreading for wetlands and uplands

Parameter	**Criteria**
pH	6-9
EC, mmhos/cm	<12
SAR	NA
CEC, meq/100g	NA
ESP, %	NA
Metals	**mg/kg**
Arsenic	40
Barium	40,000
Cadmium	10
Chromium	500
Lead	500
Mercury	10
Selenium	10
Silver	10
Zinc	500
Organics	
Oil & Grease %	NA
Flash Point, °F	>140

GUIDELINES FOR PIT OPERATIONS

Typically, successful closure of drilling waste fluid pits can be assured if pit wastes are not contaminated with extraneous materials during drilling operations. In the past, operators have used drilling pits to dispose of chemicals, waste oils, and solid wastes generated during drilling operations. Freeman and Deuel (1984) concluded that pit closure costs could be reduced if guidelines were implemented by operators to assure drilling pits were not contaminated before, during, and after the drilling operations.

Liquid Waste Management

Produced water, well test fluids, waste chemicals, and used oils are the problematic substances placed in pits for disposal. These wastes can create major problems for operators who propose closing pits to meet GLC values. Even small quantities of salty or organic wastes contaminate pit wastes to the extent on-site management cannot be utilized. For example, at an upland site where dilution burial was suggested, the GLC value for Electrical Conductivity (EC) in the waste/soil mix is /12 mmhos/cm (approximately 7,800 mg/l Total Dissolved Solids) Table 7-7. Discharging high salinity produced water into this fresh water drilling pit contaminated the waste fluids and required off-site disposal of the pit wastes by the operator.

Draining waste lubrication oil from rig engines or disposing of waste chemicals into the pit creates closure problems. Disposing of waste oil or chemicals into the pit assures costly closure procedures. Operators should implement operational guidelines to assure drilling personnel do not place contaminated waste liquids into the pit. Produced water and well test fluids should be stored in tanks and not discharged to the reserve pit. Waste oils should be collected in containers by drilling contractors and transported to recycling plants or recycled into crude oil stock tanks. Install skimmers on drainage ditches surrounding rig structures to route wash-down water, rainwater, and hydrocarbon leaks to a pit to prevent hydrocarbons from being discharged into the reserve pit. Finally, operators should evaluate their drilling mud programs to determine if pit waste volumes can be reduced.

Solid Waste Management

Empty oil and chemical drums, paint cans, paper sacks, domestic trash, and other assorted waste items are generated as solid wastes at drilling locations. In the past, many operators have placed these solid wastes in drilling pits for burial.

Implementing operational guidelines assures proper disposal of solid wastes. Storage receptacles should be placed at the drill site. Personnel should place empty paint cans, domestic trash, garbage, cable, etc., into the receptacles for proper disposal. Vendors should be required to pick up empty oil and chemical drums. Prohibiting disposal of waste solids or waste liquids into pits will assure closure procedures are less complex and costly.

PIT CLOSURE TECHNIQUES

Historical Considerations

Pit closure techniques utilized by operators in the past were relatively simple. The technique typically involved pushing levees into pits on top of wastes and compacting the closure surface area. Sometimes operators removed liquids, sometimes they did not. Surface damage has resulted at a number of locations because produced water was systematically disposed on-site in earthen pits. Hydrocarbons were buried at many sites with no possibility of significant biodegradation due to lack of oxygen; i.e., buried too deep. The technique left wastes buried in concentrated form and subject to leaching from periodic rainfall events. The technique called "out of sight, out of mind," was standard operating procedure, and is still legal in some states.

History shows burying wastes without reducing chemical constituents to harmless levels creates environmental problems. This has been supported over the past decade as abandoned sites have been discovered holding hazardous buried wastes for many years. As a result, Congress passed RCRA in 1976 to assure that industry and government disposes of waste properly.

As previously mentioned, drilling muds and brines are exempted from federal RCRA regulation, but EPA was mandated to determine if these wastes could cause harmful effects to the environment if not managed properly. Recently, states have begun to regulate waste disposal practices utilized by operators. State regulations address waste disposal techniques in various ways. Louisiana Office of Conservation (LOC) promulgated Rule 29-B, Section XV on January 20, 1986. This rule specifically addresses site restoration and is probably the most stringent in the country. Disposal of pit liquids prior to closure is regulated. Land treatment, dilution burial, solidification, or off-site disposal are pit closure options controlled under Rule 29-B for disposing of solids.

Sampling

Proper disposal of E&P wastes begins with representative samples and accurate analytical characterization. A good sampling program accurately identifies the matrix composition by:

- component or phases encountered
- volume of various components
- limiting constituents

Details on sampling large facility impoundments and remote field pits are presented in Chapter 4.

Pit Liquid Disposal

When drilling operations cease, the operator must dispose of pit liquids prior to pit closure. Otherwise, liquids will be discharged to adjacent property during pit backfilling, and in many cases, damage land or enter water courses. The two most common techniques for disposing of liquid wastes are flocculation/discharge and underground injection.

Flocculation/Discharge

Louisiana and Mississippi allow operators to discharge liquids from pits to soil, if permission is obtained from the agency. States require pit liquids to meet designated water quality criteria prior to discharge to waters of the U. S. To reduce suspended solids and precipitate metals, the pit water is treated with flocculation agents and pH adjustment chemicals, respectively. Upon successful treatment verified by laboratory tests, the water is discharged onto adjacent land or into waters of the U.S. after receiving permission from the state agency. Treatment costs can vary from $500 - $5,000, depending upon the size of the pit and its contaminated condition. An operator must weigh the economic cost of this technique compared with underground injection or off-site disposal. Some states, such as Alabama, Louisiana, New Mexico, Oklahoma, and Texas do not permit discharges of treated drilling pit water into waters of the U.S.

Underground Injection

Water and waste muds in pits can be mixed and injected into dry holes or between surface and production casing in some states. This disposal technique can significantly reduce pit closure design and costs. Mixing pit water and waste mud together for subsurface disposal reduces the volume of waste fluids managed on-site.

The procedure is relatively simple. First, permission must be obtained from the state agency before injection commences. Upon approval from the agency, earth moving equipment is used to mix and force waste liquids to one side of the pit. The liquids are picked up by a high-pressure pump and injected between the surface pipe and/or intermediate casings. Injection is continued until the pump will no longer pick up the remaining thickened mud slurry, or the remaining waste fluids in the pit must be managed on-site or removed for off-site disposal at a commercial facility.

Underground injection is an environmentally sound disposal method and helps to mitigate the impact of land surface management. Operators should assure maximum injection pressures are not exceeded. This will ensure waste fluids are not injected into an underground source of drinking water via pressure initiated fractures.

References

American Petroleum Institute (API), 1995. Metals Criteria for Land Management of Exploration and Production Wastes. Technical Support Document for API Recommend Guidance Values. API Publication Number 4600, Washington, DC, January.

Freeman, B.D. and L.E. Deuel. 1984. Guidelines for Closing Drilling Waste Fluid Pits in Wetland and Upland Areas. Presented at the 7th Annual Energy Sources Technology Conference and Exhibition. Sponsored by The Petroleum Division - ASME, New Orleans, LA, February 12-16.

8

CLEAN CLOSURE

PIT SOLIDS DISPOSAL

Land Treatment

Land treatment, which renders waste materials harmless through soil incorporation, uses dilution, chemical alteration, and biodegradation mechanisms to reduce constituents to acceptable soil levels consistent with intended land use. The technique provides both treatment and final disposal of salts, petroleum hydrocarbon (oil & grease), and metals. This waste disposal technique has proven viable by many years of use is recognized as an acceptable method to dispose of municipal sewage sludges and industrial wastes, EPA (1980).

The predominate materials in drilling fluid waste include bentonite clay and barite, Freeman and Deuel (1984). These materials are commonly found in surface soils. Accordingly, land treatment becomes an attractive disposal technique for drilling waste fluids. Drilling wastes constituents vary from well-to-well due to differences in mud programs. Analysis of these wastes must be made for each well to evaluate variable constituent levels. Constituent concentrations are compared to concentrations for salts, metals, and organics (Guidelines for Limiting Constituents, Chapter 7). This comparison identifies the land limiting constituent and dictates the type of closure procedures to assure successful land treatment design.

Freeman and Deuel (1984) discuss in detail land treatment design procedures in Appendix V. A decision diagram, Figures 8-1, 8-2 and 8-3. helps the reader formulate procedures for the land treatment of drilling waste.

Waste constituent dilution to harmless levels in soil represents the best management of drilling wastes. However, this technique is management intensive. The operator must assure the contractor follows the instructions detailed in the pit closure design. Otherwise the treatment may not be successful. Land treatment of very saline or strongly acidic and/or alkaline wastes is not recommended at upland sites, unless operators are prepared for the intensive management. Otherwise, serious damage to surface soil can result. If damage occurs, long-term restoration methods must be implemented to repair the site. This assures significantly higher closure costs.

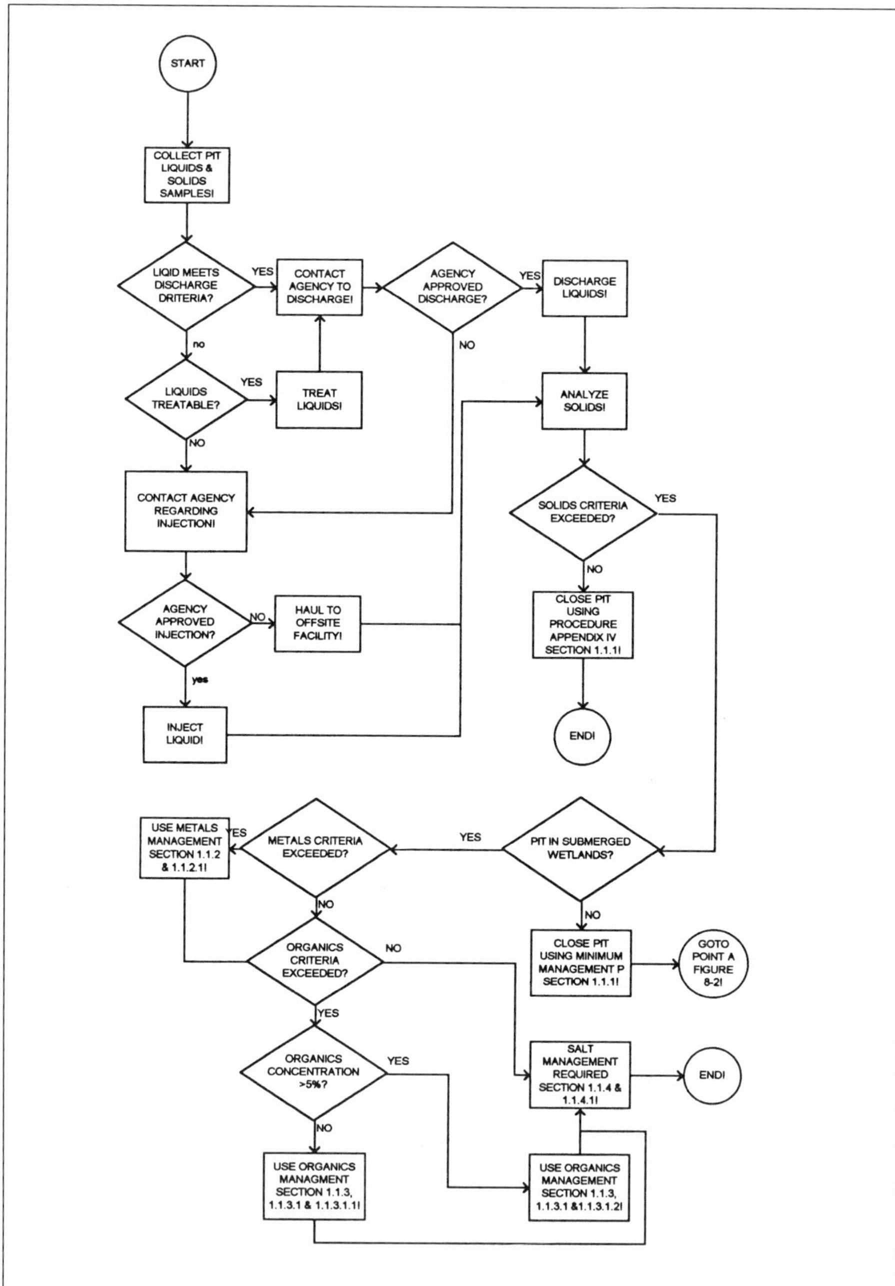

Fig. 8-1 Pit closure design diagram for use in Appendix V

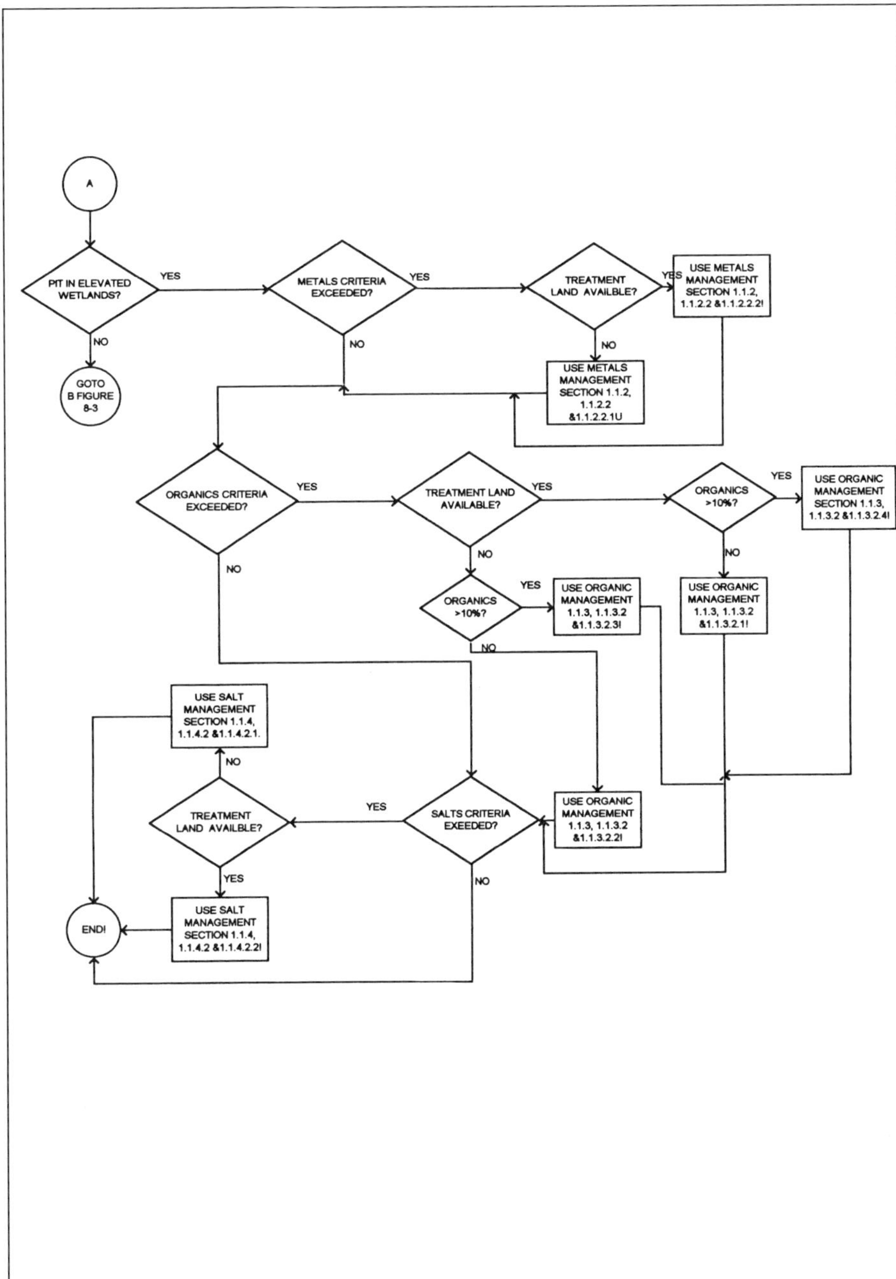

Fig. 8-2 Pit closure design diagram for use in Appendix V

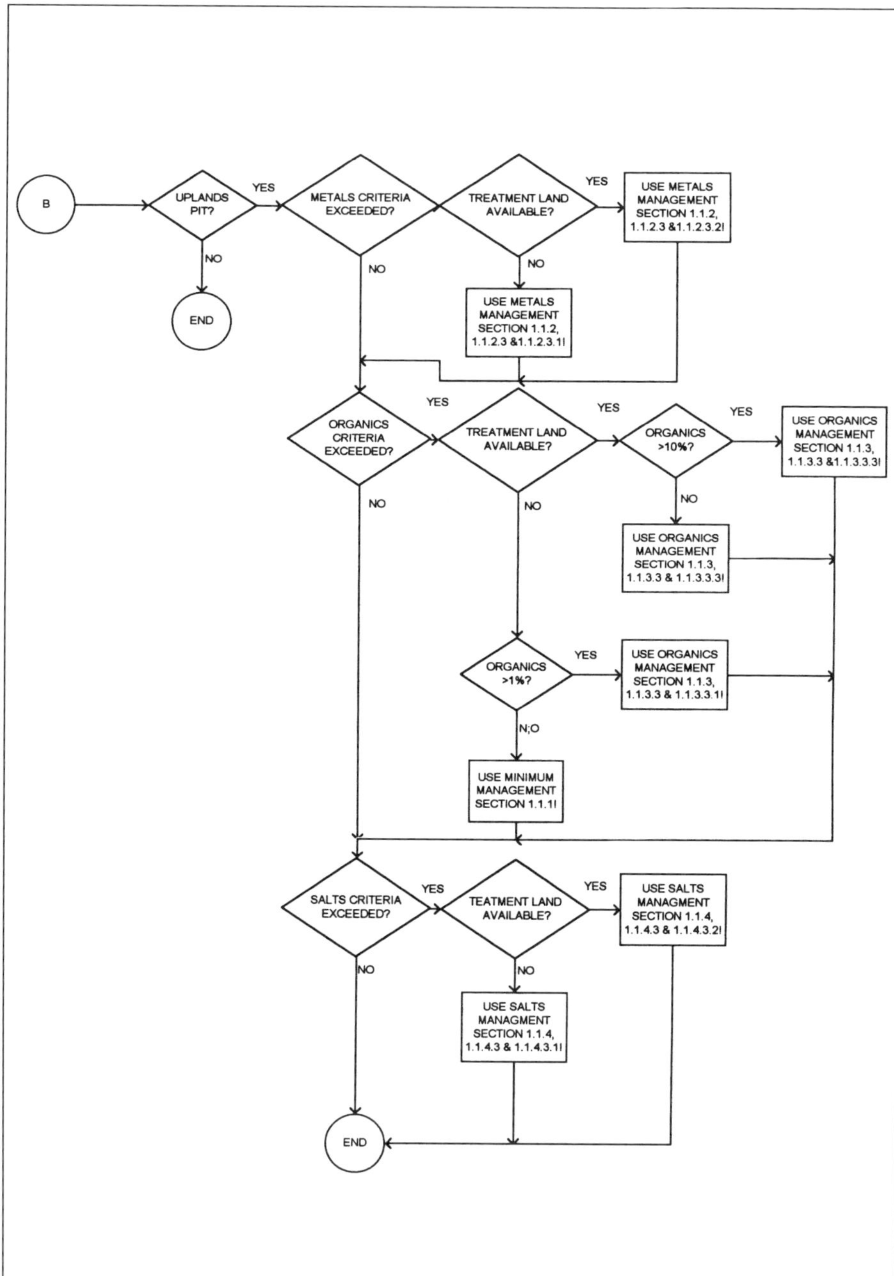

Fig. 8-3 Pit closure design diagram for use in Appendix V

The Louisiana Office of Conservation [LOC] issued regulations under Rule 29-B for closing drilling and production pits in wetland and upland areas, January 20, 1986. GLC values [Chapter 7] for land treatment are basically the same as those used by Freeman and Deuel (1984). For soluble salts, however, LOC placed stricter limitations on pits closed at elevated sites in freshwater wetlands and at upland sites. The GLC values for EC and SAR were set by LOC at the lower range of 4 and 12, respectively, Rule 29-B (1986). LOC allows operators to use higher EC and SAR values if the operators can show these increases are compatible with native soil values. The 29-B rule, amended October, 1990, allows higher barium loading, Chapter 7, and changes barium analytical methodology, Appendix II.

Dilution Burial

Dilution burial utilizes adjacent soil mixed with wastes to reduce constituent levels below GLC values. The resultant waste/soil mix is buried in trenches having at least five feet of undisturbed soil above ground water and five feet of uncontaminated soil cover over the mixed waste. Dilution and chemical alteration mechanisms are utilized in this technique to manage waste constituents. However, organic biodegradation mechanisms are restricted due to an anoxic environment. Accordingly, unless oil and grease concentration in the waste is diluted with soil to no more than 3 percent by weight before burial, dilution burial should not be used.

Dilution burial requires minimal land area for treatment, and the GLC values are less stringent with respect to total salt loadings. The technique is favored in agronomic land resource areas since roots do not penetrate the waste amended soil. Accordingly, SAR and ESP values are not applicable in dilution burial.

Dilution burial is presented as an alternative disposal technique to land treatment at upland sites. As previously mentioned, land treatment of saline wastes (EC >8) is not recommended at upland sites unless operators are prepared to conduct intensive management.

For dilution burial, ground water depths at upland sites should be at least 20 feet below the land surface. The bottom of the burial trench should be at least five feet above ground water and the top of the waste/soil mix should be five feet or more beneath the land surface. Economics dictate at least 10 feet of trench depth for storage to make the disposal technique viable. Determining the depth to ground water is imperative for dilution burial. A relatively simple technique for characterizing the subsurface moisture profile and ground water conditions is given in Appendix V .

LOC set GLC values for dilution burial, Chapter 7. These values are basically the same as land treatment with few exceptions. LOC added a

GLC value for waste moisture. Waste must not contain more than 50 percent moisture by weight prior to burial of the waste/soil mix. This requirement presents problems for operators because moisture in many clay soils already exceed the 50 percent value. In addition, the EC of the waste/soil mix must not exceed 12 mmhos/cm under LOC rules. This is equivalent to approximately 7,800 ppm Total Dissolved Solids (TDS) or about 3,900 ppm chlorides. Approximately 83 percent of freshwater base muds have EC's less than 12 (Table 8-1). Consequently, this disposal technique is used extensively at upland sites. Leachate from wastes with salt concentrations less than 7,800 ppm TDS should have minimal impacts upon ground water five feet below the burial trench. Further, the cation exchange capacity (CEC) for 70% of waste is > 15 (Table 8-2), and provides an attenuating action to reduce salt mobility. The GLC value for oil and grease must be < 3 percent for waste/soil mixtures prior to burial. No studies show the 3 percent value is not viable. Overcash and Pal (1979) show oil and grease concentrations > 1 percent in surface soils have detrimental effects on vegetation. However, this is not applicable for buried E&P wastes.

Table 8-1 Frequency Distribution for EC in Mud Pits.

EC interval mmhos/cm	Frequency Distribution	Percent of Total	Cumulative Percent
≤ 2.0	18	22.2	22.2
2.0 - 4.0	18	22.2	44.4
4.1 - 8.0	22	27.2	71.6
8.1 -12.0	9	11.1	82.7
12.1 -24.0	7	8.6	91.3
>24.0	7	8.6	100.0
Total	81	100.0	

Table 8-2 Frequency Distribution for CEC in Mud Pits.

EC interval mmhos/cm	Frequency Distribution	Percent of Total	Cumulative Percent
≤15.0	24	29.6	29.6
15.1-25.0	16	19.8	49.4
25.1-100.0	41	50.6	100.0
TotaL	81	100.0	

Road spreading

Road spreading is a disposal technique suitable for E&P wastes high in oil & grease but of moderate salt content. The technique entails a blending of E&P wastes with road oil or road mixes and spread as road material. Wastes can be incorporated directly into road bed material.

The primary limitation for road spreading is salt. A final mix or treated road surface should yield a saturated paste EC < 12 mmhos/cm. Oils should be stable within the matrix and not partitioned as a separate component by water since contaminated storm water may require a storm water permit.

E&P wastes containing weathered crude with a large asphaltic fraction are ideally suited for disposal via road spreading. An E&P waste with a flash point < 140 °F should not be disposed of using a road spreading technique because of the fire hazard.

Solidification

Solidification and stabilization are waste fixation processes which are intended to immobilize, encapsulate, or contain waste constituents in a solidified form to prevent contamination of soil and ground water. Solidification and stabilization are separate treatment techniques, accomplishing the same end. Solidification of wastes involves the production of a solid mass having sufficiently high structural integrity to allow transport and/or disposal of the solid without requiring secondary containment. Stabilization involves immobilization of constituents in wastes by chemical alteration to form insoluble compounds, or by entrapment within the solidified product. Major types of fixation include cementitious/pozzolanic processes which envelopes the waste solids in a materials matrix.

Solidification is a viable disposal option but is more expensive than land treatment or dilution burial. Operators may find this option practical and cost effective. However, complying with test requirements under Rule 29-B may be difficult. The operator must demonstrate the buried solid has integrity by complying with specific values related to compressibility, wet-dry cycling, permeability, and leachate tests (Table 8-3). These tests are conducted on the final solidification product to demonstrate material performance or compliance. We consider these tests necessary to eliminate mobility of contaminates.

Off Site Disposal

There are drilling locations where injection, land treatment, dilution burial, road spreading or solidification techniques can not be used. The only waste disposal option remaining is hauling the waste to a commercial facility for proper disposal. This may present problems for operators in some states, since commercial waste disposers may not be available. Therefore, the wastes must either be hauled to a hazardous waste disposal facility or hauled long distances to out-of-state commercial disposal sites.

Presently, this is not a major problem since only Louisiana and Oklahoma require that pits be closed using specified techniques.

Off site disposal is the most expensive waste disposal alternative. The operator must pay to remove the wastes from the pit, pay to transport wastes to the disposal site, and pay to dispose of wastes at the commercial site. This costly disposal technique gives the operator incentive to "manage" drilling pits or to use alternative mud systems that minimize waste volumes.

Table 8-3 Solidification Performance Standards

Chemical Parameters	**Criteria (1:4 Leachate*)**
pH	6-12 s.u.
Chlorides	<500.0 mg/l
Oil & Grease	<10.0 mg/l
Heavy Metals	**Criteria (Leachate Testing*)**
Arsenic	<0.5 mg/l
Barium	<10.0 mg/l
Cadmium	<0.1 mg/l
Chromium	<0.5 mg/l
Lead	<0.5 mg/l
Mercury	<0.02 mg/l
Selenium	<0.1 mg/l
Silver	<0.5 mg/l
Zinc	<5.0 mg/l
Physical Parameters	**Criteria (Testing**)**
Unconfined Compressive Strength	>20 lbs/in^2
Permeability	<$1x10^{-6}$ cm/sec
Wet/Dry Durability	>10 cycles to failure

* *Leachate testing methods are given in the Laboratory Manual for the Analysis of Oilfield Waste (Department of Natural Resources, August 9, 1988, or latest revision.*

** *Testing must be conducted according to ASTM or other approved methods prior to pit closure by solidification processes.*

Conclusions

All waste pits require some simple closure procedure to assure protection of soil and ground water resources. Simple closure requires blending waste materials with berm or levee soil. Any other type closure design should be considered alternative management and requires more detailed closure specifications.

Analytical tests by EPA and API show E&P waste solids are non-hazardous. Accordingly, chemicals, production fluids, waste oils, trash and other debris should not be placed in the pit. Expedite closure to eliminate others from utilizing the pit.

References

EPA, 1980. Hazardous waste land treatment, SW-874. September 1980.

Freeman, B.D. and L.E. Deuel. 1984. Guidelines for closing drilling waste fluid pits in wetland and upland areas. Presented at the 7th Annual Energy Sources Technology Conference and Exhibition. Sponsored by The Petroleum Division - ASME, New Orleans, Louisiana, February 12-16, 1984.

Rule 29-B. 1990. Section XV. - Pollution control, State of Louisiana, Department of Natural Resources. Office of Conservation, Baton Rouge, Louisiana, Promulgated January 20, 1986, revised October 20, 1990.

9

SOIL AMENDMENTS

INTRODUCTION

Extensive research sponsored by EPA and the oil and gas industry demonstrates the vast majority of E&P wastes are not hazardous. However, most E&P wastes require management to prevent adverse impacts to land and water.

Excessively alkaline reactions, salts, and the dispersed nature of drilling waste solids (lack of physical structure) represent the principal limitations associated with these wastes. Some wastes contain high hydrocarbon levels, while other wastes may contain high levels of total metals.

SOIL REACTION AMENDMENTS

Use of aluminum sulfate ($Al_2(SO_4)_3$) controls excess alkalinity in pit wastes. Excess acidity is managed with lime (CaO). Amendments applied to waste liquids will not react properly with solids unless blended by mechanical action.

Final pH adjustments of waste solids are made after spreading on land surfaces, and mixing with surface soils at upland sites. Using agricultural limestone ($CaCO_3$) adjusts excess acidity and elemental sulfur (S) adjusts excess alkalinity. Final pH adjustments are made after homogenizing pit contents, but prior to mixing with the levee materials at wetland sites.

For details on desirable soil reaction targets and the various mechanisms involved refer to Appendix III.

SALINITY AMENDMENTS

Only a rinse or leach process with fresh water reduces salts in soils and E&P waste solids. Poor physical condition and inherently low hydraulic conductivity prevents free exchange of salt without the aid of structuring amendments, wetting agents and/or mechanical processes.

Structuring amendments aid by increasing soil aggregation. These agents help prevent dispersion of the clays at recent produced water spills by allowing water to infiltrate. This aids chemical amendments in entering the soil matrix. Polymer products increase soil aggregation. These agents include polyvinyl alcohols, polyacrylamides and natural plant polymers. Mechanical processes physically loosen the soil allowing a greater surface area for soil contact with water and chemical amendments. Disking can loosen the soil to a depth of 6-10 inches depending on the equipment available. Chiseling the soil is necessary on sites where salt contamination occurs at greater depths.

Sodicity Amendments

The primary salt associated with oil field wastes or produced waters is sodium chloride (NaCl). Soils, drill cuttings and other E&P waste solids exposed to high salt levels naturally become sodium saturated or sodic. Calcium amendments counteract sodicity and high salinity. If the material is both saline and sodic, calcium should be introduced prior to the leach process. The amendment of choice is dependent upon solid phase calcium levels and available time.

Solid Phase Calcium in Excess

This is a condition typified by produced waters spilled to calcareous soils, resulting in a sodic soil condition, or where sodic drilling fluids are applied to calcareous soils. The amendment of choice is elemental sulfur, when time is not a consideration, and mechanical means are available to distribute the amendment. The sulfur oxidizes to sulfuric acid. This reacts with calcium carbonate ($CaCO_3$)to form calcium sulfate ($CaSO_4$).

When time is a factor or mechanical means are not available for distribution, soluble calcium sources, such as calcium chloride ($CaCl_2$) or calcium nitrate ($Ca(NO_3$) are necessary. Also, soluble calcium salts are effective when soil or waste solids have not yet equilibrated to a sodic condition, (i.e., applied immediately after a spill). Agricultural grade gypsum is an effective calcium amendment when time is not a factor and mechanical means are available for distribution. Gypsum is readily available at low cost.

ORGANIC MATTER AMENDMENTS

Organic amendments in the form of peat moss or low salt manures improve soil structure. Also organic materials serve as a sink for salts by exchange mechanisms and dilute salts in soil solution by increasing moisture holding capacity.

BIOREMEDIATION AMENDMENTS

Native soils generally contain a sufficient number and species diversity in microflora (bugs) to degrade petroleum hydrocarbon under controlled applications. In rare cases, inoculations may be necessary to offset poor population dynamics. Typically, an inordinately long lag phase results from limited populations. Addition of commercial microbes reduces the lag phase. However, their effectiveness is difficult to measure when compared to regulatory benchmarks such as oil and grease (O&G) and total petroleum hydrocarbons (TPH).

Fertility adjustments in the form of fertilizer amendments, aeration and moisture are necessary for bioremediation program. Fertilizer amendments include nitrogen (N), phosphorous (P) and potassium (K). Ammonium nitrate (NH_4NO_3), ammonium sulfate ($(NH_4)_2SO_4$) and urea are good nitrogen sources. They should be applied in multiple applications to increase efficiency, because nitrogen is lost to the atmosphere with time. Nitrogen should be disked-in to distribute the application and aerate the system.

Nitrogen is applied at rate yielding a carbon:nitrogen (C:N) ratio of 150:1. Phosphorous and potassium are applied at 1/4 the rate of nitrogen (600:1). Superphosphate and muriate of potash are suitable sources. Phosphorous and potassium are applied in a single application with the first nitrogen treatment.

Moisture content should be maintained within 50-80 % of field capacity to effect optimum degradation rates, e.g., about 1 in of water per week.

E&P SPILL EVENTS AND TREATMENT SCENARIOS

The following scenarios illustrate the calculations and procedures required to remediate typical oil and gas field contaminated soils.

Scenario I
Light Crude Oil Spill on Soil
Oil spill onto native soil, 20 bbl of West Texas crude oil gravity 20-30° API.

A. Considerations

Oil spilled on native soil spreads in the direction of the surface drainage with pooling in depressions. Vertical infiltration is controlled by:

- surface condition
- soil porosity
- pore size distribution
- soil moisture
- soil structure
- viscosity and gravity of the oil

The most important factors affecting infiltration are permeability of the soil profile, surface soil condition, and soil moisture content. The amount of oil percolating into a soil profile is controlled by the permeability of the least pervious horizon or layer of soil in the vertical cross section.

Texas Railroad Commission (Statewide Rule 91) requires removal of all pooled oil and immediate delineation of the area of contamination, both vertically and horizontally. Area of contamination means the affected area with more than 1% total petroleum hydrocarbon (TPH). This procedure applies equally well world wide and can be applied at all oil spill locations.

Any soil containing more than 5% (TPH) light crude requires removal or reduction to 5% or less by mixing with uncontaminated background soil (Tables 9-1 and 9-2). In Louisiana and Texas, all contaminated soil must be reduced to 1% or less TPH within 1 year following the spill event. Typically, bioremediation is not effective below 6-10 inches in depth. Accordingly, contaminated soil below that depth must be excavated and treated on the surface.

B. Scenario Site Assessment

Crude oil with an API gravity of 20 has a specific gravity of 0.93. A 20 bbl spill of this oil yielding an average concentration of 1% oil to a depth of 6-in. will occupy an area of 14,100 ft^2 (0.33 acre-6 in). The same volume yielding an average concentration of 5% oil in soil affects 2,820 ft^2 (20 bbl × 42 gal/bbl × 8.3 lb/gal × 0.93)/(2,000,000 lb/acre-6 in × 0.05) = 0.065 acre-6 in or 2,249 ft^2 (0.065 acre-6 in × 34,560 ft^2 /acre).

Stained soil defines the surface boundaries of a recent oil spill. Samples collected in vertical profile define concentrations with depth. Unless specific information dictates otherwise, profiles are sampled initially in 0-6, 6-12, 12-24 and 24-36 inches in depth intervals. Profiles should be taken at various points along a transect which defines surface drainage of the site. Sample the site in grid fashion where drainage patterns and internal percolation are not well defined.

C. Remediation

The following remediation assumes a 5% TPH level in a 6 in surface layer following the removal of pooled oil. Contaminated soil below the treatment zone is considered to be ≤1% TPH.

This loading rate assumes a mean annual soil temperature of 22°C or higher. Decrease allowed TPH concentration in the soil to be treated, 1% for each 7° C decrease in temperature below 22°C (Table 9-1).

Table 9-1 Initial TPH loading rates for oil; 20° API and above

Mean Annual Soil Temperature, oC	Loading Rate, % TPH
>22.0	5
15.0-21.9	4
8-14.9	3
<8	0

Microbial degradation of petroleum hydrocarbon does not occur when soil temperatures remain below 5°C. Consider composting operations when soil temperatures remain below 10°C. This involves removing TPH contaminated soil and incorporating it in a pile of organic base material such as high protein hay and animal manure. The heat generated by decomposition of the organic material elevated the treating temperature.

Remediation entails biodegradation of crude oil using native soil microflora. Parameters having a profound affect on rates of degradation include:

- initial oil level (loading rate)
- carbon:nitrogen ratio and other fertilizer inputs
- moisture content
- aeration

These parameters can be controlled in the field.

Research shows the optimum carbon:nitrogen (C:N) ratio for field degradation of petroleum hydrocarbon is 150:1. A lower C:N ratio (more fertilizer) theoretically will improve degradation rates. Generally some other variable such as water use efficiency or free air exchange limits the utility of adding more nitrogen. Nitrogenous compounds released by Kjeldahl digests add little available nitrogen reserves, and should not be considered in determining the C:N ratio. Nitrate-N and ammonium-N released by extraction with a potassium chloride solution comprise the available nitrogen fraction and must be considered in determining the C:N ratio, if the field or spill area has been fertilized recently with nitrogen.

Phosphorous and potassium are elements that may be limiting and cause a positive response when added to the system. A nitrogen: phosphorous: potassium (N:P:K) ratio of 4:1:1 provides a good balance of these elements.

Mechanically till fertilizer applications into the soil to distribute the amendment through the treatment zone. Tilling also infuses oxygen into the soil. Split N applications coupled with tillage operations increase fertilizer use efficiency, and lessens the potential for eutrophic stagnation (the creation of an oxygen deficit by nutrient enrichment).

Moisture levels should be maintained between 50-80% of the soil water holding capacity (field capacity). This requires addition of irrigation water (about 1 in/week) in dry climates (arid and semiarid). Humid regions with spurious rainfall distribution patterns may require seasonal irrigations to overcome drought conditions.

D. Calculations

Assumptions:

- Soil temperature > 22° C, TPH loading rate = 5%
- 1 ac-6 in = 2,000,000 lb (91.8 lb/ft^3)
- ppm, ([mg/kg]) x 2 = lb/ac.-6 in
- crude oil contains 78% carbon (C)

1) carbon (C) content, in ppm (mg/kg), of soil due to oil spill

$$\begin{aligned} C,\% &= TPH\% \times 0.78 \\ C,\% &= 5.0\% \times 0.78 \\ C,\% &= 3.9\% \\ C,\ mg/kg &= 3.9\% \times 10{,}000 \\ &= 39{,}000\ mg/kg \end{aligned}$$

2) nitrogen (N) requirement

$$C/N = 150/1$$
$$(39{,}000\ mg/kg)/N = 150/1$$
$$N = 260\ mg/kg$$
$$N = (260\ mg/kg) \times 2$$
$$N = 520\ lb/ac\text{-}6\ in$$

3) spill site occupies 0.06 ac-6 in

$$N = 520\ lb/ac\text{-}6\ in \times 0.06\ ac\text{-}6\ in$$
$$N = 31\ lb$$

4) NPK 4:1:1

$$P{,}K = 31\ lb\ N/4$$
$$31/4 = P{,}K = 8\ lb$$

E. Procedure

1. Remove all pooled oil
2. Excavate contaminated soil below 6 in, (e.g., >1% TPH.)
3. Mix in place with background soil to 5% by weight TPH. Use the EC equation from Chapter 1 for determining the amount of background soil required. Maintain the treatment area depth of mixed soil ≤6 in
4. Add half of the ammonium nitrate(NH_4NO_3), 47 lb (33% N), 87 lb superphosphate (9.2 % P), and 16 lb muriate of potash (51% K)
5. Disk and cross disk to 6 in to distribute fertilizer and aerate treatment zone
6. After 6 to 8 weeks add half of the remaining ammonium nitrate (NH_4NO_3), 23 lb, disk and cross disk to distribute fertilizer and aerate treatment zone
7. After 6 to 8 weeks add the remaining ammonium nitrate, 23 lb, disk and cross disk to distribute fertilizer and aerate treatment zone
8. Irrigate as necessary to maintain treatment zone moisture between 50 and 80% of field capacity
9. After 6 to 8 weeks sample site to demonstrate cleanup level <1% TPH. Retreat as above if necessary

Scenario II
Heavy Crude Oil Spill on Soil
Spill on to native soil, 20 bbl of heavy crude oil, gravity 10-20° API.

A. Considerations

Heavier crudes are not as amenable to biodegradation as the lighter weight crudes. To compensate, reduce the initial petroleum hydrocarbon loading rate for 10-20° API crude to 3% TPH. The temperature considerations for TPH loading rates are shown in Table 9-2 .

Table 9-2 Initial TPH loading rates for oil ; 20° API and below

Mean Annual Soil Temperature, oC	Loading Rate, % TPH
>22.0	3
15.0-21.9	2
8-14.9	1
<8	0

Microbial degradation of oil does not occur below 8°C soil temperature. However, composting is possible. Add an equivalent amount of non-petroleum hydrocarbon, (e.g.; peat moss or cellulose fiber), as an extender and adsorbent. The added carbon sequesters; the more labile TPH fractions and added nutrients. This provides a more favorable soil environment for colonization of indigenous microflora and proliferation of oil "eaters". Animal manure may also be used. Fresh manure requires more tillage operations to maintain an aerobic system, since animal manures contain low C:N ratios and degrade rapidly.

B. Site Assessment

Crude oil with an API gravity of 10° has a specific gravity of 1. A 20 bbl spill having an average concentration of 10% oil to a depth of 4 in will occupy an area approximately 2,294 ft^2 (48 ft x 48 ft).

Mix the contaminated soil with fresh (background) soil to achieve the recommended 3% loading rate. The following procedure demonstrates one way of controlling the application rate, while limiting the treatment zone to the biologically active surface. A treatment zone depth of 5 in is selected for calculation and design purposes. This depth of incorporation and mixing is readily achieved using conventional farm equipment, (e.g., rubber tired tractor and disk).

The proper loading rate is achieved by a combination of removal and fill operations which result in a total treatment area (inclusive of the original spill boundary) of 6,098 ft2 (127 ft x 48 ft).

C. Calculations

Assumptions:

- soil temperature > 22°C, TPH loading rate = 3%
- acre-6 in = 2,000,000 lb (91.83 lb/ft3)
- oil equally distributed in 4 in spill area profile at 10%
- background soil oil residual 0.01%
- 5 in depth of mixing
- crude oil contains 78% carbon (C)

1) contaminated soil mass

$$\begin{aligned} 20\text{ bbl} \times 42\text{ gal/bbl} \times 3.8\text{ l/gal} &= 3{,}192\text{ l} \\ 3{,}192\text{ l} \times 1\text{ kg/l} \times 2.2\text{ lb/kg} &= 7{,}022\text{ lb oil} \\ (7{,}022\text{ lb oil}/\text{ X lb soil})100 &= 10\%\text{ oil in soil} \\ X &= 70{,}220\text{ lb soil} \end{aligned}$$

2) background soil requirement for controlled loading to 3%

$$\begin{aligned} \text{TPH, }3\% &= [(10\%)(70{,}220\text{ lb})+(0.01\%) \\ &\quad (\text{X lb soil})]/(70{,}220\text{ lb}+\text{X lb soil}) \\ 210{,}660\text{ lb}+3X &= 702{,}200 + 0.01X \\ X &= 164{,}395\text{ lb} \\ \text{Treatment area mass} &= 164{,}395\text{ lb} + 70{,}220\text{ lb} \\ &= 234{,}615\text{ lb} \end{aligned}$$

$$\begin{aligned} \text{area equivalent} &= 234{,}615\text{ lb} \;/\; 1{,}666{,}667\text{ lb/ac-5 in} \\ &= 0.14\text{ ac} \\ &= 0.14\text{ ac} \times 43{,}560\text{ ft}^2/\text{ac} \\ &= 6{,}098\text{ ft}^2\ (48\text{ ft} \times 127\text{ ft}) \end{aligned}$$

3) carbon (C) content in ppm (mg/kg)

$$\begin{aligned} \text{C, \%} &= \text{TPH,\%} \times 0.78 \\ \text{C, \%} &= 3.0\% \times 0.78 \\ \text{C, \%} &= 2.3\% \\ \text{C, mg/kg} &= 2.3\% \times 10{,}000 \\ \text{C, mg/kg} &= 23{,}000\text{ mg/kg} \end{aligned}$$

4) nitrogen (N) requirement

$$C/N = 150/1$$
$$(23{,}000\ mg/kg)/N = 150/1$$
$$N = 153\ mg/kg$$
$$N = 153\ mg/kg \times 2$$
$$N = 306\ lb/ac\text{-}6\ in$$
$$N = 255\ lb/ac\text{-}5\ in$$

5) treatment area occupies 0.14 ac

$$N = 255\ lb/ac\text{-}5\ in \times 0.14\ ac$$
$$N = 36\ lb$$

6) N:P:K = 4:1:1

$$N/4 = P,K = 36/4$$
$$P,K = 9\ lb.$$

D. Procedure

1. Remove all pooled oil
2. Excavate 1.5-in soil from receiving area adjacent to spill and stockpile in holding area (Fig. 9-1)
3. Remove 2.5-in from spill area to receiving (treatment) area. Spread and level to grade (Fig. 9-1)
4. Backfill spill area with uncontaminated soil stockpiled in holding area. Spread and level to grade (Fig. 9-1)
5. Add half of the ammonium nitrate (NH_4NO_3), 55 lb (33% N), all of the super phosphate, 98 lb (9.2% P), and all of the muriate of potash, 18 lb (51%) to the spill and receiving areas
6. Disk and cross disk to 5-in to distribute fertilizer and aerate treatment zone
7. After 6 to 8 weeks add half of the remaining ammonium nitrate (NH_4NO_3); (27 lb), disk and cross disk to distribute fertilizer and aerate treatment zone
8. After 6 to 8 weeks add final 27 lb ammonium nitrate, disk and cross disk to distribute fertilizer and aerate treatment zone
9. Irrigate as necessary to maintain treatment zone moisture between 50 and 80% of field capacity, about 1-in of water per week
10. After 6 to 8 weeks sample site to demonstrate cleanup level <1% TPH. Retreat as above if necessary

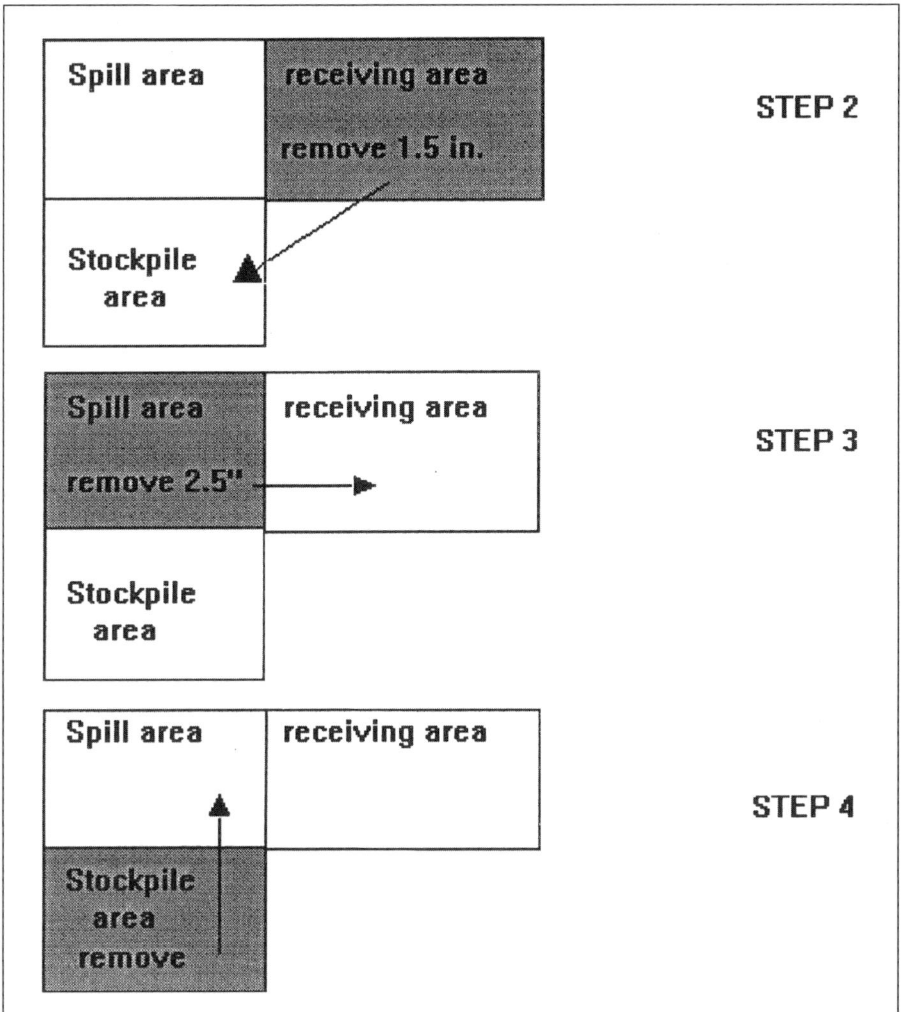

Fig. 9-1 Soil moving plan for Scenario II

Scenario III
Light Crude Oil and Produced Water Spill on Soil

West Texas crude oil plus produced water were spilled onto native soil. The mixture is 40 bbl of produced water with 10 bbl of oil.

A. Considerations

Oil distribution from a spill event involving mixtures of oil and water are controlled by the amount of water, infiltration capacity, and soils capacity to transmit water through the profile. Typically, infiltration capacities are higher than percolation rates. This creates a condition of surface saturation and causes spreading of contamination during spill events.

Produced waters typically are high in sodium chloride (NaCl). A typical oil field produced water from West Texas is high in TDS (Table 9-3).

Table 9-3 Properties of a produced water from West Texas

Parameter	Value
pH, s.u.	6.8
EC, mmhos/cm	251
Na, mg/l	42,869
Ca, mg/l	13,848
Mg, mg/l	2,437
Cl, mg/l	88,147
SO_4, mg/l	991
HCO_3, mg/l	327

Specific conductance (EC) is related to the total salt in solution (total dissolved solids, TDS) by the following approximate relation,

$$TDS = EC \times 613$$

where 613 is a constant determined empirically for oil field produced waters. Summation of ions (Table 9-3) yields a TDS of 148,619 mg/l, comparing well to the calculated value of 153,863 mg/l (251 x 613). Forty barrels (40 bbl) produced water at 153,863 mg/l TDS is equivalent to 2,161 lb salt [(153,863 mg/l) × (3.8 l/gal) × (42 gal/bbl) × (2.2 lb/kg) × (kg/10^6 mg.) × 40 bbl]. The impact of this much salt spilled on the land is profoundly influenced by the area, and eventually the matrix volume of the receiving soil.

EC evaluated from the spill area saturated paste provides the best measure of produced water impact. Saturated paste EC values for the spill area and receiving soil texture are shown in Table 9-4. Calculations are based on the affected soil receiving 0.06, 0.6 and 1.2-in brine/acre, percolating to depth of 6-in.

Table 9-4 The effects of salt and texture on saturated paste EC

Total Salt, lb	Spill Area, acres	Texture	SP Moisture*, %	SP EC, mmhos/cm
2,161	0.05	sand	28	127
2,161	0.05	loam	38	93
2,161	0.05	clay	48	74
2,161	0.1	sand	28	63
2,161	0.1	loam	38	47
2,161	0.1	clay	48	37
2,161	1.0	sand	28	6.3
2,161	1.0	loam	38	4.7
2,161	1.0	clay	48	3.7

** These saturated paste moisture percentages represent satisfactory average values for calculation purposes.*

On-site treatment options for restoring the land to full agronomic potential include rinse and fill, *in situ* leach, dilution burial and landspreading techniques.

Oil floating on the water is assumed to affect the same spatial volume as the produced water. The loading is approximately 3.4% oil in soil for a 10 bbl oil spill onto 0.05 ac. Corresponding oil loadings for 0.1 ac is 1.7 %, and 1 ac is 0.2%. TPH levels of 3.4% and 1.7% require treatment to 1% within one year under Railroad Commission of Texas (RRC) Statewide Rule 91.

B. Remediation Procedures

The five recommended remediation methods are known as minimum management, rinse and fill, in situ leach, dilution burial, and landspreading.

Minimum Management Method

Description. EC values between 8 and 16 mmhos/cm significantly reduce agronomic use of the land, Total agronomic loss occurs when EC > 30

mmhos/cm. Given the small acreage involved, it may be more prudent to abandon the land as wasted and plant non-agronomic halophytic shrub seedlings (fourwing saltbush, *Atriplex canescens*; winterfat, *Ceratoides lanata*; prostrate kochia, *Kochia prostrata*) to stabilize the landscape. This is considered a minimum management and is not intended to restore agronomic potential.

Calculations. Fertilizer calculations and application rates for Minimum Management Procedure are

1) carbon (C) content in ppm (mg/kg)

$$C,\% = TPH,\% \times 0.78$$
$$C,\% = 3.4\% \times 0.78$$
$$C,\% = 2.65\%$$
$$C, mg/kg = 2.65\% \times 10{,}000$$
$$C, mg/kg = 26{,}500 \; mg/kg$$

2) nitrogen (N) requirement

$$C/N = 150/1$$
$$(26{,}500 \; mg/kg)/N = 150/1$$
$$N = 177 \; mg/kg$$
$$N = 177 \; mg/kg \times 2$$
$$N = 354 \; lb/acre\text{-}6 \; in$$

3) treatment area occupies 0.05 ac

$$N = 354 \; lb/ac\text{-}6 \; in \times 0.05 \; ac$$
$$N = 18 \; lb$$

4) NPK 4:1:1

$$N/4 = P,K = 18 \; lb/4$$
$$P,K = 5 \; lb$$

To obtain the necessary K and P concentrations use 60 lb of 10-10-10 fertilizer. This contribution provides 6.0 lb of N (60 × 0.10), 2.6 lb P (6.0 × 0.44), and 5 lb K (6.0 × 0.83) (See Chapter 1). Supplement the P with 26 lb [(5.0 - 2.6)/ 0.092] of super phosphate. For the first application add all of the P and K and half of the total N (9 lb). Six pounds (6 lb) N are added with the 10-10-10 fertilizer so only 3.0 lb N need to be added as ammonium nitrate(NH_4NO_3) (33% N).

$$\text{Ammonium Nitrate } (NH_4NO_3) = (3.0 \; lb/0.33) = 9.1 \; lb$$

Procedure.

1) Apply 2.5 tons good quality hay or cellulose fiber to spill area
2) Apply 9.1 lb ammonium nitrate (NH_4NO_3), 60 lb of 10-10-10 fertilizer, and 26 lb super phosphate to spill area
3) Disk and cross disk to 6 in to distribute organic matter and fertilizer
4) After a 6 to 8 week interval apply 14 lb [(9/2)/0.33] ammonium nitrate(NH_4NO_3); (33% N) and disk to distribute fertilizer and aerate soil

After a 6 to 8 week interval apply remaining 14 lb ammonium nitrate and disk to distribute fertilizer and aerate soil

6) Irrigate as necessary to maintain soil between 50-80% of field capacity
7) After 6 to 8 weeks from last fertilizer application sample and test to demonstrate TPH < 1%. Repeat treatment if necessary
8) Plant seedling halophytes on 2 m (6.6 ft) centers

Rinse and Fill Method

Description. This technique is used for low volume, high brine, coarse or readily flocculated soils. Soils are excavated, suspended in a high calcium solution (0.1N calcium chloride, 3.6 lb $CaCl_2$/bbl) and flocculated with a centrifuge operation. Solids having EC's between 8 to 16 mmhos/cm are returned to the excavation and mixed with an equal volume of undisturbed soil to complete salt amelioration. Solids with EC>16 are reflocculated. Hydrocarbon is biodegraded to complete site restoration if not removed during the rinse operation.

Procedure.

1) Remove salt contaminated soil (40 yd^3 for the 0.05 acre spill area) to a large mixing container
2) Blend soil with 0.1N $CaCl_2$ solution to suspend solids and displace brine
3) Centrifuge to separate solids from rinsate
4) Separate oil from rinsate and dispose of the latter by well injection
5) Return solids having <16 mmhos/cm EC to excavated area, blending with an equal volume of native soil. Blend by disking to 6 inches

6) Add 200 lb high quality hay or cellulose fiber to increase aeration and organic matter. Add 50 lb 10-10-10 fertilizer and disk to 6 inches. Seed site with native grasses
7) Test for petroleum hydrocarbon to verify TPH <1%. Repeat, if necessary

In situ Leach Method

Description. *In situ* leach is applicable when a spill occurs on deep well drained soils, with low-salt leach-water available and no threat exists to usable groundwater. Artificial drainage to a collection sump for disposal can be used to remove salts from shallow soils, or those with an impeding layer. Artificial drainage can also be used to protect groundwater resources by intercepting brackish leachate. The soil profile from the surface downward must be sufficiently permeable for drainage to occur.

The spacing between drain lines will be determined by the hydraulic conductivity and depth to restrictive layer or desired rooting profile. As a general rule, the spacing of laterals ranges from 15 to 60 ft. One and a half pore volumes of water are required to remove excess salt from the profile. The amount of water to be drained and the available time are also important design considerations. Installing drains 2 to 3 ft deep increases the number of laterals, but decreases the volume of water necessary to flush salt. Do not bury drains more than twice the depth of the surface layer in stratified soil (i.e., coarse textured surface over finer textured subsurface layer).

Gypsum application rates can be determined either from sodium adsorption ratio (SAR) as it relates to the exchangeable sodium percentage (ESP), or it may be calculated from actual ESP measurements. The latter is usually employed under non-equilibrium conditions, (i.e., soil recently contaminated).

Calculations. *Calculation of SAR:* Calculation is from the produced water (Table 9-3). Convert to meq/liter from mg/liter: Meq/liter is equal to mg/liter / meq wt.

Sodium, Na+	42,869 mg/liter / 23 mg/meq	=	1,864 meq/liter
Calcium, Ca++	13,848 mg/liter / 20 mg/meq	=	692 meq/liter
Magnesium, Mg++	2,437 mg/liter / 12 mg/meq	=	203 meq/liter

$$SAR = \text{Soluble Na(meq/l)}/[((\text{Soluble Ca(meq/l)} + \text{Soluble Mg (meq/l)})/2)]^{1/2}$$
$$SAR = 1{,}864 \ / \ [(692 + 203)/2] = 88$$

Calculation of ESP from SAR.

ESP = [100(-0.0216 + 0.01475 × SAR)] / [1+(-0.0216+0.01475 × SAR)]
ESP = [100(-0.0216 + 0.01475 × 88)] / [1+(-0.0216+0.01475 × 88)] = 56%

This calculation indicates a soil in contact with oilfield produced water having an SAR of 88 will equilibrate to an ESP of 56%.
Gypsum requirement determined from CEC and ESP: It is undesirable for the ESP to exceed 12% of the CEC. To compensate, Ca^{+2} is added in the form of gypsum at a rate equivalent to the actual exchangeable sodium minus the desired exchangeable sodium. Gypsum requirement will vary dependent on soil type and corresponding CEC (Table 9-5).

$$\frac{(\text{Actual \% ESP-Desired \% ESP})}{100} \times \text{CEC} = \text{CR (charge requirement)}$$

$$\frac{(\text{Actual \% ESP- 12\%})}{100} \times \text{CEC} = \text{CR (charge requirement) (Tons/acre-ft)}$$

Divide by 2 for determining tons per acre 6″.

This following example is based on data for sand (Table 9-5).

$$\frac{(\text{56\% ESP- 12\% ESP})}{100} \times (\text{3.5 meq/100g CEC}) = \text{1.54 (charge requirement)}$$

Gypsum requirement = 1.54 × 1.7 (Tons $CaSO_4$ / acre ft)
= 2.6 (Tons $CaSO_4$ / acre ft

Table 9-5 Gypsum requirements for various soil types

Texture	CEC meq/100g	SAR	% ESP	Gypsum Requirement lb/.05 ac-ft
Sand	3.5	88	56	262
Loam	15.0	88	56	1,122
Clay	35.0	88	56	2,618

Procedure.

1) Install drain lines 2 to 3 ft deep at spacing designed to remove the desired quantity of water within a given time interval
2) Laterals are connected to a common line draining to a sump. Drain lines are covered with rock or gravel followed by soil and contoured to grade
3) Add gypsum at the application rates calculated (Table 9-5)

4) Apply 50 lb of 10-10-10 fertilizer and disk to distribute and aerate
5) Wet soil with water and maintain between 50 and 80% of field capacity during biodegradation phase
6) After 6 to 8 weeks apply 15 lb ammonium nitrate and disk to distribute and aerate
7) After 6 to 8 weeks apply 15 lb ammonium nitrate and disk to distribute and aerate
8) Chisel soil to 16 inches following a 6 to 8 week reaction interval
9) Impound salt affected soil with a levee and flood irrigate with 4 inches of fresh water (5,430 gal/0.05 acres)
10) After water penetrates surface flood again with 4 in to effect salt removal
11) Leachate collected in sump should be commingled with oil field produced water for well injection

Dilution Burial Method

Description. Dilution burial applied to salt and hydrocarbon contaminated soil is designed to reduce salinity to EC <24 mmhos/cm and hydrocarbon levels to <1% TPH. Groundwater depth should be at least 40 ft below land surface for an EC of 24 mmhos/cm, and 20 ft for an EC threshold of 12 mmhos/cm. The bottom of the burial trench should be at least 25 ft above groundwater for an EC of 24 mmhos/cm and 5 ft for an EC of 12 mmhos/cm. The top of the contaminated/clean soil mix should be 5 ft below the land surface.

Assumptions:

spill area	=	0.05 ac
depth of penetration	=	6 in
depth to groundwater	>	40 ft

values to use in calculations:

Soil Value	EC, mmhos/cm	ESP, %	TPH, %
Spill Area	127	25.0	3.2
Background	0.8	1.0	<0.01
Target Threshold	24	12	1

Calculations: limiting constituent

1. *contaminated soil, volume*

$$\text{Soil Volume} = 0.05 \text{ ac} \times 43{,}560 \text{ ft}^2/\text{ac} \times 0.5 \text{ ft}$$
$$= 1{,}089 \text{ ft}^3$$

2. soil volume requirement for salinity management

$$24 \text{ EC} = [(127 \text{ EC})(1{,}089 \text{ ft}^3) + (X \text{ ft}^3)(0.8 \text{ EC})]/(1{,}089 \text{ ft}^3 + X \text{ ft}^3)$$
$$X = 4{,}835 \text{ ft}^3)$$

3. soil volume requirement for ESP Management

$$12 \text{ ESP} = [(25 \text{ ESP})(1{,}089 \text{ ft}^3) + (X \text{ ft}^3)(1.0 \text{ ESP})]/(1{,}089 \text{ ft}^3 + X \text{ ft}^3)$$
$$X = 4{,}026 \text{ ft}^3$$

4. soil volume requirement for organic management

$$1 \text{ TPH} = \frac{(3.2TPH)(1089 \text{ ft}^3) + (X \text{ ft}^3)(0.01TPH)}{(1089 \text{ ft}^3 + X \text{ ft}^3)}$$

$$X = 2{,}420 \text{ ft}^3$$

Salinity, as shown by the above calculations, is the limiting constituent. The salinity dilution requirement is 4,835 ft^3. This plus the contaminated soil volume of 1,089 ft^3 yields a total trench volume of 5,924 ft^3. A 15 ft trench depth is required for 10 ft of storage depth and 5 ft for cover. A trench 3 ft wide by 197 ft long will accommodate 5,910 ft^3 of contaminated/clean soil mix. Installing several trenches, totaling 197 ft, in an array surrounding the spill area expedites reclamation. For example, we assumed the 0.05 acre spill is circular and 4 trenches (49.25 ft each) are opened to receive the contaminated soil/native soil mix (Fig. 9-2). No treatment for ESP is required. However, we recommend a blanket application of gypsum be made based upon soil type (Table 9-6).

Table 9-6 Blanket gypsum applications for various soil types

Texture	Gypsum Requirement lb/acre
Sand	150
Loam	500
Clay	1,500

Procedure.

1) Open trench A (Fig. 9-2) 49.25 ft long x 15 ft deep x 3 ft wide, segregating the initial 6.8 ft of material removed for use as backfill, topsoil cover
2) Excavate contaminated soil area 1 (1,089/4 = 272.25 ft^3) and

place in trench A blending mechanically with 1,208.75 ft^3 (4,835/4) of uncontaminated mix soil

3) Repeat steps 1 and 2 for trenches B, C, and D and excavation areas 2, 3 and 4, respectively
4) Backfill trench areas with cover material, placing the deeper excavated materials in first, followed in sequence with topsoil replaced last
5) Remaining topsoil is used as backfill in the spill area
6) Apply 50 lb/ac 13-13-13 fertilizer to the disturbed area. Area may be seeded or allowed to revegetate naturally

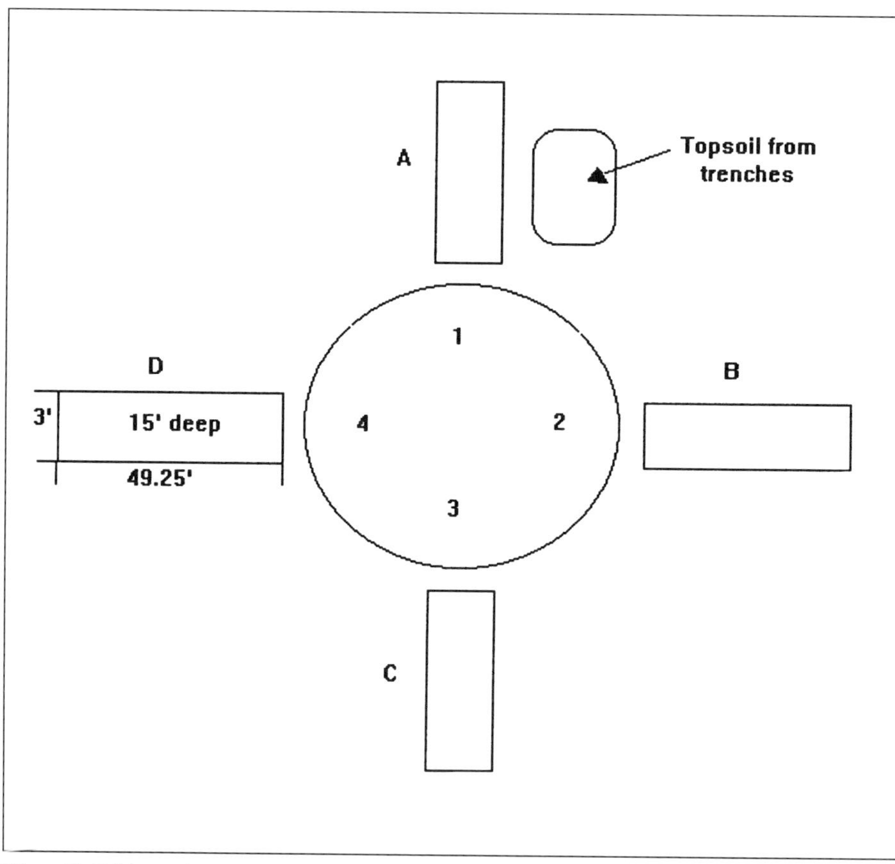

Fig. 9-2 Trench construction schematic

Land Treatment Methods

Description. Land treatment is a technique for rendering E&P waste solids and contaminated soils harmless through soil incorporation. These methods use dilution, chemical alteration, physical adsorption, and biodegradation processes in reducing constituents to acceptable levels consistent with intended land use.

Method No. 1

Method 1 Assumptions:

spill area	=	0.05 acre
depth of penetration	=	6 in
depth to groundwater	>	40 ft

Values to use in calculations:

Soil Value	EC, mmhos/cm	TPH, %
Spill Area	127	3.2
Background	0.8	<0.01
Target Threshold	8	1

Method 1 calculations: limiting constituent

1) *contaminated soil, volume*

Soil Volume = $0.05\ \text{ac} \times 43{,}560\ \text{ft}^2/\text{ac} \times 0.5\ \text{ft}$
= $1{,}089\ \text{ft}^3$

2) *soil volume requirement* for salinity management

$8\ \text{EC} = [(127\ \text{EC})(1{,}089\ \text{ft}^3) + (X\ \text{ft}^3)(0.8\ \text{EC})]/(1{,}089\ \text{ft}^3 + X\ \text{ft}^3)$

$X = 17{,}999\ \text{ft}^3$

3) *soil volume requirement* for organic management

$1\ \text{TPH} = [(3.2\ \text{TPH})(1{,}089\ \text{ft}^3) + (X\ \text{ft}^3)(0.01\ \text{TPH})[/\ (1{,}089\ \text{ft}^3 + X\ \text{ft}^3)$

$X = 2{,}420\ \text{ft}^3$

Salinity, as shown by the above calculations, is the limiting constituent. The dilution requirement is 17,999 ft^3. To incorporate to six inches requires; 0.88 ac of land. 0.88 ac = 19,087 ft^3 / 21,780 ft^3/ac 6 in.

Method 1 Procedure:

1) Scrape contaminated soil from the spill area and spread evenly over 0.88 ac to a depth of 0.3 in. Strongly structured and massive soils will be difficult to spread thinly and may require a thicker initial application on less acreage followed by redistribution over 0.88 acres

2) Disk the treatment area to a depth of 6 inches
3) Apply a blanket application of 136 lb gypsum ($CaSO_4$) to the treatment area and disk again
4) The excavated spill area is restored to the original contour with treated soil
5) Apply 75 lb ammonium nitrate and 50 lb 13-13-13 fertilizer to the disturbed area. Area may be seeded or allowed to revegetate naturally

Method No. 2.

Method 2 Assumptions:

spill area	=	0.1 ac
depth of penetration	=	6 in
depth to groundwater	>	40 ft

Values to use in calculations:

Soil Value	EC, mmhos/cm	TPH, %
Spill Area	63	1.7
Background	0.8	<0.01
Target Threshold	8	1

Method 2 Calculations—limiting constituents:

a. Contaminated soil, volume

Soil Volume = 0.1 ac × 43,560 ft²/ac × 0.5 ft
= 2,178 ft³

$$\text{Soil Volume} = 0.1\text{ ac} \times 43{,}560\text{ ft}^2/\text{ac} \times 0.5\text{ ft} = 2{,}178\text{ ft}^3$$

b. Soil volume requirement for salinity management

$$8\text{ EC} = [(63\text{ EC})(2{,}178\text{ ft}^3) + (X\text{ ft}^3)(0.8\text{ EC})]/(2{,}178\text{ ft}^3 + X\text{ ft}^3)$$

$$X = 16{,}638\text{ ft}^3$$

c. Soil volume requirement for organic management

$$1\text{ TPH} = [(1.7\text{ TPH})(2{,}178\text{ ft}^3) + (X\text{ ft}^3)(0.01\text{ TPH})]/\ (2178\text{ ft}^3 + X\text{ ft}^3)$$

$$X = 1{,}540\text{ ft}^3$$

Salinity, as shown in the above calculations, is the limiting constituent. The dilution requirement is 16,638 ft^3. To incorporate to 6 in requires 0.76 ac of land.

Method 2 Procedure:

1) Scrape contaminated soil from the spill area and spread evenly over 0.76 ac to a depth of 0.40 in. Strongly structured and massive soils will be difficult to spread thinly and may require a thicker initial application on less acreage followed by redistribution over 0.76 acres
2) Disk the treatment area to a depth of 6 in
3) Apply a blanket application of 136 lb gypsum ($CaSO_4$) to the treatment area and disk again
4) Restore the excavated spill areas to the original contour with treated soil
5) Apply 75 lb ammonium nitrate and 50 lb 13-13-13 fertilizer to the disturbed area. Area may be seeded or allowed to revegetate naturally

VEGETATIVE RESTORATION

Objective

The final step to site closure is establishing a vegetative cover. A vegetative cover serves two purposes: (1) erosion control, and (2) aesthetically pleasing landscape. The latter is important since it instills a degree of confidence that no significant levels of contamination linger on-site.

Tillage Requirements

Grass seeds are small and some effort is required to establish an adequate seedbed. Disk and harrow the seedbed to fracture the soil into sufficiently small aggregates so as to provide good seed and soil contact. Tillage of the soil is controlled primarily by the moisture level. If the soil is worked wet, particularly when mixing already dispersed materials, the soil becomes puddled resulting in a near structureless condition. If worked too dry, clods may be formed that are not readily fractured.

Seeding

Grass seed should be broadcast on roughened surfaces and rolled. Germination and seedling survival is improved with a hay mulch cover. Do not work seed into soil deeply. The planting depth should not exceed 1/2 in and is best if it can be controlled between 1/8 and 1/4 in.

Fertility Requirements

The seeding formulations given in Table 9-7 are designed for quick ground cover and a certain degree of permanence. Permanence is controlled ultimately by the fertility status of the growth media. Typically, native soils have a low innate fertility relative to nitrogen, phosphorous, and potassium. Accordingly, fertilizer is required to ensure adequate fertility.

We recommend fertilizing with a minimum 300 lb/acre application of 10-10-10 or 12-12-12 during the tillage operation. Adding clover to the seed formulations aids maintenance of nitrogen fertility. However, a top-dress application of 30 lb of nitrogen may be required 6-8 weeks following emergence to ensure the clover survives.

Table 9-7 West Texas plant species seeding rates

Season	Plant Species	Seeding Rate, lb./acre
Spring/ Summer	Common bermuda grass	10
	Kleingrass-75	20
	Weeping lovegrass	5
	Clare subterranean clover	8
Fall/ Winter	Common bermuda grass	10
	Kleingrass-75	20
	Weeping lovegrass	5
	Ryegrass	40
	Clare subterranean clover	8

10

NATURALLY OCCURRING RADIOACTIVE MATERIALS

INTRODUCTION

Naturally Occurring Radioactive Materials (NORM) result from the decay of uranium 238 and thorium 232 located within the earth's crust. Forty (40) daughter products are found. Radium-226 is the most common isotope found in NORM. NORM appears as an ingredient of scale on water wet surfaces of oil field equipment, (i.e., separator dump valves, heater treater dump valves, tubing and produced water tanks). Additionally, NORM in the form of lead 210 bearing scale appears in propane pumps. Experience shows NORM is almost always contained in sulfate scales. However, NORM can be present in carbonate scales. NORM typically is a source of "alpha" radiation particles. Alpha particles have low power and do not penetrate paper or skin; however, when inhaled, Alpha particles can damage the human body. Accordingly, the most effective protection from NORM is good personal hygiene and breathing protection. Currently, Louisiana, Michigan, Mississippi, New Mexico, Texas, and Utah have adopted regulations dealing with NORM; however Oklahoma is considering regulating NORM.

LOUISIANA NORM RULE

Introduction

While Louisiana is not the only state controlling NORM, the Louisiana rule (1995) covers all of the essential protection subjects. The rule applies to all persons generating, extracting, processing, possessing, using, treating, transporting or disposing of NORM. Additionally, the rule applies where NORM and the radiation emitted are not considered advantageous to the materials or products. Also, the rule applies to Technology Enhanced Natural Radioactive Material (TENR): meaning natural sources of radiation which could not normally expose persons or the environment to radiation without some technological activity (drilling, producing) not expressly designed to produce radiation.

Exemptions

NORM wastes and NORM contaminated materials (soil) are exempted from the rule, if the NORM concentration is:

- ≤5 picocurie/g (ρci/g) of radium-226 or radium-228 above background, or
- ≤150 ρci/g of any other NORM radionuclide.

Equipment containing NORM is exempt from the rule if the radiation exposure ≤50 μrem/hr.

Land areas are exempt if samples taken from 100 m^2 are:

- ≤5 ρci/g of radium 226 or 228 above background average within the top 15 cm. and ≤15 ρci/g above background averaged over each subsequent 15 cm., or
- ≤30 ρci/g of radium 226 or 228 averaged over 15 cm increments of depth, provided the total effective dose equivalent does not exceed 0.1 rem/yr.

Radiation Surveys

The survey must rely on a calibrated instrument. Upon completing the survey of contaminated facilities, sites, equipment, or excavated contaminated land, the operator must develop and submit for approval written procedures to ensure worker protection, Appendix VI. On-site maintenance will be authorized only if the maximum radiation is ≤2 μrem/hr above background at any point in the work area. NORM concentrations exceeding the rule threshold require the submission of a confirmatory survey report. The survey report must include the qualifications of the instrument operator.

General License

A general license can be issued for persons receiving, possessing, owning, storing or transferring NORM not exempt from the rule. A copy of the confirmatory survey (Fig. 10-1) must accompany the notification. Additionally, each license must prepare and submit for approval written procedures for surveying sites and equipment to assure NORM is not released into areas not controlled by the rule (unrestricted areas), unless the site or equipment has been decontaminated.

Storage

NORM may be stored by a licensee no more than 90 days from date of generation. After 90 days, the NORM waste must be transferred to an authorized facility for treatment, storage or disposal.

Contaminated surface equipment, excluding wellhead, removed from service, must be decontaminated to ≤50 μrem/hr or disposed of within one year of the date the equipment was removed from service.

Pipe Storage and Production Equipment Yard (General)

License. Yards receiving, possessing, processing and cleaning tubulars or equipment contaminated with NORM scale or residue having ≤50 μrem/hr are covered by a General License, if:

- a written program is submitted outlining procedures for screening incoming shipments to assure they do not exceed the 50 μrem/hr threshold
- a written worker protection program is submitted for approval (Appendix VI)
- a written worker protection program is submitted for approval regarding controlling oil contamination
- a written program is submitted for approval regarding preventing NORM from escaping beyond the facility boundaries
- before releasing the site for unrestricted use, decontaminate the site to ≤5 ρci/g of radium-226 or -228

A specific license is required for yards decontaminating tubulars or equipment exceeding the 50 μrem/hr limit.

NORM Disposal. NORM must be transferred to an authorized land disposal facility or disposed of by alternate methods authorized in writing by the Agency upon application by the operator.

For NORM containing NOW at concentrations ≤30 ρci/g of radium 226 or 228, the waste can be transferred to an approved commercial facility for treatment, if:

- radium-226 or -228 after treatment dilution does not exceed 5 ρci/g above background, and

- the receiving facility screens incoming waste for NORM and radium-226 or -228 does not exceed 30 ρci/g and,
- agency approves, or
- for NOW where 5 ρci/g NORM (200 ρci/g specifically licensed facilities under Appendix VII can accept the waste

Also, melting of NORM contaminated scrap may be authorized by specific license, if end product or melt by-product contains ≤5 ρci/g radium 226 or 228.

Storage Containers. A NORM storage container must be closed at all times, except when necessary to add or remove waste. At least quarterly NORM storage containers must be inspected and damaged containers replaced. A record of the inspection must be made and retained for 5 years. All NORM containers must be labeled and stacked so each label is visible from the access isle or area.

NORM Manifesting. A shipment of NORM controlled by rule must be accompanied by a manifest (Figure 10-2 and -3). The manifest must designate the recipient, e.g., an approved treatment, storage or disposal site. The manifest must include sufficient copies so a copy can be retained by each transporter, storer and disposer. Additionally, one copy must be retained by the shipper. The shipper must provide a description of the NORM being shipped and guidelines for controlling emergency situations involving the waste. The shipper must receive confirmation of acceptability of the proposed waste prior to shipment. The recipient must report any discrepancies between the shipped and received NORM within 15 days of receipt of the waste.

NORM DISPOSAL

Where legally allowed, NORM, like any waste, should be retained by the generator. Any time waste is disposed at a contract facility, the contractor becomes the custodian of the waste. If any problem arises at the contract facility, the contractor looks to the generator for corrective action. Accordingly, the generator should gain approval from the state to mix NORM scale with cement and pump the mixture into a well during plugging operations. NORM can be injected (Texas statewide Rule 94). Contaminated tubing or line pipe can be cemented into the hole during abandonment.

Facility/Location: ____________________ Date: ________

Meter Model: ____________________ Serial No.: ________

Detector Type: [ ] Model Serial No.: ________

[ ] Model Serial No.: ________

Battery Check [ ] Source Check [ ]

Calibration Date: ________

Source Type: ________

Background Radiation Level: ________ microR/hr

Description of Equipment/Material Surveyed: ____________________

Item/Material Surveyed (Description, Serial Number, Size Quantity, etc.)	Maximum microR/hr.

Comments: ____________________

Survey(s) Conducted By: ____________________
(Print Name)

____
(Signature)

Fig. 10-1 Example of a NORM survey data sheet

STRAIGHT BILL OF LADING—SHORT FORM—NOT NEGOTIABLE—ORIGINAL

RECEIVED, subject to the classifications and tariffs in effect on the date of the issue of this Bill of Lading.

EP-2038 (5-82)

CARRIER: I. M. TRUCKER	PAGE 1 OF 1 PARTS	SHIPPER'S NO 149019
FROM: OUR WAREHOUSE, INC.	REFER TO NO.	DATE 02/03/XX
AT: 123 Storage Lane, Anywhere, USA	OUR ORDER NO.	CUSTOMER'S ORDER NO

CONSIGNED TO (mail or street address for notification only)

NORM Storage and Disposal, Inc.
123 Scenic Highway
Anywhere, USA

Our Warehouse Inc.
(Signature of Consignor)

☐ PREPAID

COLLECT ☐

ROUTE	ROUTE CODE	CAR/TRUCK NO.	SEAL NUMBERS

NO. & KIND OF PACKAGES: DRUMS	PAILS	CASES	DESCRIPTION OF ARTICLES (Show D.O.T. Hazardous Materials First)	WEIGHT	CLASS OR RATE
25			Radioactive Material, Low Specific Activity, n.o.s.,	15,000 lbs.	
			Radioactive Material, UN2912 (226 Ra and 228 Ra),		
			Normal Form, Barium Sulfate scale containing Ra nuclides,		
			< 0.0075 microCuries/gram, Radioactive White-I		

TOTAL NO. OF PACKAGES	PALLET WEIGHT	DUNNAGE WEIGHT	TOTAL WEIGHT
25			15,000 lbs.

COD — COLLECT ON DELIVERY $ — REMIT TO: OUR WAREHOUSE, INC. — C.O.D. Charge to be paid by: SHIPPER ☐ CONSIGNEE ☐

MAIL FREIGHT BILL IN DUPLICATE WITH NO. 3 COPY OF BILL OF LADING TO:

FOR PRODUCT EMERGENCY Spill, Leak, Fire, Exposure or Accident CALL CHEMTREC – DAY OR NIGHT 800-424-9300

SHIPMENTS VIA MOTOR CARRIER

D.O.T. HAZARDOUS MATERIALS PLACARDS FURNISHED BY ☐ SHIPPER ☐ CARRIER

* If the shipment moves between two ports by a carrier by water, the law requires that the bill of lading shall state whether it is carrier's or shipper's weight.

OUR WAREHOUSE, INC. SHIPPER — CARRIER

PER I. M. Shipper — PER — AGENT

1

MAIL TO CONSIGNEE OR AS INSTRUCTED

Fig. 10-2 Example of a Manifest for onshore shipment of materials containing NORM

SHIPPING PAPER / DANGEROUS CARGO MANIFEST

No. 001

SHIPPER NAME: OFFSHORE OIL COMPANY
AND LOCATION: OFFSHORE PLATFORM A
SHIP TO: OUR WAREHOUSE, INC.

VESSEL NAME: BO TRUC 27
VESSEL NO. / CALL SIGN: [illegible]
VESSEL NATIONALITY: USA

RADIOACTIVE MATERIALS

PROPER SHIPPING NAME	HAZARD CLASS	I.D. NO.	RADIONUCLIDE	FORM	ACTIVITY	STOWAGE LOCATION	NO. / TYPE PACKAGES	GROSS WEIGHT
RADIOACTIVE MATERIAL, LSA, n.o.s.	RADIOACTIVE MATERIAL	UN2912	(228 Ra AND 226 Ra)	NORMAL FORM	< 0.0075 microCuries/ grams	ON DECK	4 / 55 GAL. DRUMS	800 # / DRUM

BARIUM SULFATE SCALE CONTAINING Ra NUCLIDES / EXTERNAL RADIATION LEVEL: 5 MILLIREMS / HOUR / RADIOACTIVE WHITE - I

FOR PRODUCT EMERGENCY, DAY OR NIGHT, CALL CHEMTREC AT 1-800-424-9300, TOLL FREE

CERTIFICATION: THIS IS TO CERTIFY THAT THE ABOVE-NAMED MATERIALS ARE PROPERLY CLASSIFIED, DESCRIBED, MARKED AND LABELED AND ARE IN PROPER CONDITION FOR TRANSPORTATION ACCORDING TO THE APPLICABLE REGULATIONS OF THE DEPARTMENT OF TRANSPORTATION.
IF TRANSPORTED BY HELICOPTER: THIS SHIPMENT IS WITHIN THE LIMITATIONS PRESCRIBED FOR PASSENGER AIRCRAFT / CARGO AIRCRAFT ONLY (DELETE NON-APPLICABLE)

SIGNATURE: PREPARER I. M. Operator DATE: 02/01/XX

I HEREBY DECLARE THAT THE CONTENTS OF THIS CONSIGNMENT ARE FULLY AND ACCURATELY DESCRIBED ABOVE BY PROPER SHIPPING NAME AND ARE CLASSIFIED, MARKED, AND LABELED, AND ARE IN PROPER CONDITION FOR TRANSPORT BY BOAT / HELICOPTER (DELETE NON-APPLICABLE) ACCORDING TO APPLICABLE INTERNATIONAL AND NATIONAL GOVERNMENTAL REGULATIONS.

SIGNATURE: VESSEL'S MASTER OR LICENSED OFFICER: I. M. Skipper DATE: 02/01/XX

Fig. 10-3 Example of a manifest for water shipment of dangerous cargo

References

Department of Environmental Quality, 1989. Nuclear Energy Division, State of Louisiana, *Regulation and Licensing of Naturally Occurring Radioactive Materials* (NORM), Title 33, Part XV, Chapter 14; September.

Railroad Commission of Texas, 1995. *Disposal of Oil and Gas NORM Waste*, RRC § 3.94 Rule 94, February 1.

APPENDIX I

LIMITING CONSTITUENTS FOR LAND DISPOSAL OF PETROLEUM EXTRACTION INDUSTRY WASTES

INTRODUCTION

Onshore Exploration and Production (E&P) functions generate large quantities of non-hazardous and exempt wastes. The vast majority of this volume is produced water (estimated at 98%), which, except for small quantities used for wildlife watering, are injected into Class II wells (40 CFR Part 146). The remaining volume consists of spent drilling fluids, treating fluids and associated wastes. Associated wastes include, produced sand, tank bottoms and other materials circulated from wells or separated from oil, water and gas flowlines or tank battery facilities prior to custody transfer. E&P wastes typically are disposed on-site. However, a small quantity goes to off-site facilities.

This appendix develops limiting concentrations of petroleum hydrocarbons and salts allowable for one time land treatment, burial and road-spreading of E&P wastes. Parameters providing evidence of proper saline waste management include:

- Electrical Conductivity (EC)
- Sodium Adsorption Ratio (SAR)
- Exchangeable Sodium Percentage (ESP)

and for hydrocarbons:

- Total Petroleum Hydrocarbons (TPH)

The limiting values for land application of E&P wastes apply to soil/waste mixtures. We recommend the following values for various use conditions.

Table A-I-1 Limiting Criteria For Land Treatment

	Wetlands			Uplands	
	Brackish	**Freshwater Submerged**	**Freshwater Elevated**	**Forest/ Recreational**	**Farming/ Residential**
pH (s.u.)	6-9	6-9	6-9	6-8	6-8
EC (mmhos/cm)	–	–	8-50	4-8	2-4
SAR	–	–	14-100	<14	<12
CEC (meq/100g)	>15	>15	>15	>15	>15
ESP (%)	–	–	<25	<15	<15
Heavy Metals (mg/kg)			**Wetlands**		**Uplands**
Arsenic			<10		<40
Total Barium			<20,000		<40,000
Cadmium			<10		<10
Chromium			<500		<500
Lead			<500		<500
Mercury			<10		<10
Selenium			<10		<10
Silver			<10		<10
Zinc			<500		<500
Organics (%)			**Wetlands**		**Uplands**
Oil & Grease			≤1		≤2

One-time waste application producing soil/waste mixtures having values at or less than the above limits result in minimal impact on soil and vegetation. Except for the most sensitive crops, the waste application results in less than 15% reduced yield during the first year after application. Typically, total recovery of crop yield potential occurs within 3 years. However, in some cases, soil amendments such as gypsum and nitrogen containing fertilizers may be required.

Literature Review

Investigators (Miller and Honarvar, 1975; Ferrante, 1981; Freeman and Deuel, 1984; Nelson, 1984) identified salts and hydrocarbons, found in E&P wastes, as responsible for reduced crop yields. These contaminants, in excess of given thresholds values, result in phytotoxicity. Additionally, sodium salts deteriorate soil structure (porosity and permeability) resulting in reduced plant water availability, excess water runoff and erosion. Also, salts and hydrocarbons can adversely impact surface and ground waters (Henderson, 1982; Murphy and Kehew, 1984).

SALINITY

Soil retains moisture which dissolves some soil constituents. The dissolved material enters the soil solution as cations and anions in proportion to their solubility in the matrix. Major ions observed in E&P wastes include calcium (Ca), sodium (Na), magnesium (Mg), potassium (K), chloride (Cl), sulfate (SO_4), bicarbonate (HCO_3), carbonate (CO_3) and hydroxide (OH). The more ions dissolved in the aqueous phase, the higher is the electrical conductivity of that phase. Accordingly, electrical conductivity (EC) becomes a measure of the ionic strength of the soil solution. Sodium Adsorption Ratio (SAR) and Exchangeable Sodium Percentage (ESP) measure the influence specific soluble ions (Na, Ca, Mg) impose on the soil. Accordingly, when viewed together, EC, SAR, & ESP, represent a measure of the available cations and ions in the soil solution. This combination is given the general term of salinity.

HYDROCARBONS

Hydrocarbons, when released to surface soils, penetrate to varying depths depending on soil type. Oil has an affinity for clay. Accordingly, oil does not penetrate deeply into clay soils. On the other hand, oil penetrates deeply into sandy soil.

Several measures of hydrocarbons are available. EPA measures hydrocarbons by a standardized analytical protocol known as Oil and Grease (40 CFR Part 136). This test, when applied to soils, is time consuming and costly. Since Oil and Grease (O&G) method represents the official EPA method, EPA, SW-846, Method 3540), Annex H, we use it in this appendix. However, based on extensive research (Deuel and Holliday, 1993), we recommend use of the analytical method known as Total Petroleum Hydrocarbons (TPH). Two variations of TPH are available, TPH-Infrared and TPH-GC-FID.

Total Petroleum Hydrocarbons - Infrared (TPH-IR)

This method (EPA 600/4-79-020, Method 418.1) provides good results on fresh spill of hydrocarbons to soils. However, for mature spills,

Deuel and Holliday (1993) found the method does not accurately reflect the decrease in hydrocarbon content of soil as the result of bioremediation because the method reports biomass as TPH. This results in demonstrating extended half-lives of TPH or no significant reduction as a result of treatment.

Total Petroleum Hydrocarbons - Gas Chromatography with Flame Ionization Detection (TPH-GC-FID)

TPH can be determined by Gas Chromatography (GC) and Flame Ionization Detection (FID). Deuel and Holliday (1993) confirmed by Loss on Ignition the GC-FID analytical method provided a reliable method of monitoring decrease in soil oil content as treatment progresses.

DISSOLVED CONSTITUENT RELATIONSHIPS

The soluble constituents (ions) present in soil solution conduct electrical current. The conductivity depends on type and concentration of ions. This ability to conduct electrical current is electrical conductivity (EC). EC, measured directly in water extracts of soil and solid waste (Annex B), is reported in millimhos per centimeter (mmhos/cm) which is recognized as the reciprocal of units of resistance. Dissolved salts represent the predominate ions within the soil solution. Accordingly, EC is used as an indirect measure of Total Dissolved Solids (TDS).

Barrow (1966) showed an exact correlation exists between EC and a specific salt dissolved in distilled water at low concentrations. In high concentrations, mixed salt species solutions or the presence of non-ionic dissolved components, the correlation becomes inaccurate. A statistical correlation

$$\mathrm{TDS} = \mathrm{A} \times \mathrm{EC} \qquad \text{Eq. A-I-1}$$

was developed by regression analysis. Hem (1985) found the range of "A" (slope) varied from 540 to 960 cm. mg/mmhos liter for various concentrations of dissolved ions in the presence of impurities. U.S. Salinity Laboratory staff (1954) recommended A = 640 for naturally occurring saline-sodic soils. We find for E&P soil/waste mixtures A = 613 provides a better correlation. This is based on API (1987) and EPA (1987) data review. Using the following values in Equation A-I-1:

$$EC = 4 \text{ mmhos/cm}$$
$$A = 613 \text{ cm. mg/mmhos liter}$$
$$TDS = 4 \times 613$$
$$TDS = 2452 \text{ mg/liter}$$

Hydrocarbons and fine clay particles interfere with the TDS analysis (Annex C). Accordingly, TDS is not an accurate measure of soil salinity. Therefore, we recommend determining the degree of salinity of the soil using Equation A-I-1.

DISSOLVED CONSTITUENT IMPACTS

Dissolved contaminants (hydrocarbons and ions) adversely impact plants and soils.

Plant Impacts

Hydrocarbons and salts dissolved in the soil moisture increase the osmotic pressure of the soil solution. Plant roots in contact with high osmotic pressure soil solution do not posses the ability to overcome the osmotic pressure of the soil and die from lack of water (Haywood and Wadleigh, 1949; U.S. Salinity Laboratory staff, 1954). The point of permanent wilting occurs when plants cannot recover in the presence of non-saline water. Interestingly, a statistical correlation exists between soil osmotic pressure and EC:

$$\text{Osmotic Pressure (OP), atm} = 0.36 \times EC \qquad \text{Eq. A-I-2}$$

The uptake of salts from contaminated soil disrupts nutrient uptake and utilization by antagonism (Kramer, 1969). Accordingly, excess salt alters the composition of soil solution, reducing the availability of essential plant nutrients. The result is lowered yield and crop quality. However, no one threshold salinity level exists for all plants (Maas and Hoffman, 1979). Maas (1986) shows vegetables are more sensitive to salt than grasses and grains, Plates 12, 13 and 14. Maas (1986) developed the data for Plates 12, 13 and 14 from agricultural projects receiving salt containing water during a long period of time;

single time applications of waste to soils elevate the response. Lunin (1967) suggests doubling the threshold limits associated with continual use system for single application situations.

The U.S. Salinity Laboratory staff (1954) developed a general crop response to salt, Table A-I-2. Basically, the Table suggest no effect on crop yield at or below EC = 2 and gross crop yield reduction or complete failure for all but the most salt tolerant species at EC = 16 or greater.

Table A-I-2 General crop response as a function of EC.(After U. S. Salinity Laboratory Staff, 1954)

EC, mmhos/cm	Affect on crop yield
0 - 2	None
2 - 4	Slight
4 - 8	Many crops affected
8 - 16	Only tolerant crops yield well
> 16	Only very tolerant crops yield well

If drainage is present, excess salinity can be leached from the soil by rainfall or application of irrigation water. Also, growing salt tolerant plants aids in soil recovery since the root growth allows air and water to travel through the soil (Foth and Turk, 1972). Salt-contaminated soil typically responds to application of calcium sulfate (gypsum). The response mechanism consists of replacing the exchangeable sodium ions with calcium ions (Oster and Rhoades, 1984). The presence of iron in salt contaminated soil apparently aids calcium/sodium exchange (Maas, 1986). Accordingly, for gypsiferous soils, plants tolerate EC values about 2 mmhos/cm greater than shown in Plates 12, 13 and 14.

Producing or swabbing into the reserve pit increase spent mud salinity. Miller and Pesaran (1980) observed decreased plant growth in a 1:1 mud/soil mixture when the mud contained elevated salt concentrations. Miller and Pesaran (1980) test results showed for mud/soil mixtures having EC<8:

- 7% decrease in growth for green beans
- 13% decrease in growth for corn

Nelson (1984) showed an average decrease in yield of 20% and 38% for swiss chard and rye grass, respectively, when EC ranged from 6.3 to 18.6 mmhos/cm. These results demonstrate the impact high EC values.

On the other hand, Tucker (1985) reported no adverse yield impact on Bermuda grass and alfalfa at EC values of 1.3 to 5.3 and 1.7 mmhos/cm, respectively. Tucker (1985), also, reported a significant decrease in soil EC values as a function of time following soil/waste mixing. He attributed this to leaching of the salts from the plant root zone.

For sensitive areas, wetlands, working to a soil/waste mixture EC <4 is expected to result in less than a 15% decrease in yield for most crops. In reality, since the adverse impact of EC decreases rapidly as salts leach from the soil and, most oil and gas fields are not in sensitive areas or used to raise vegetable crops, working to an EC <8 normally is adequate. Refer to Tables A-I-1 and 7-6 for specific guidance.

Leaching Impacts

In areas of wet surface infiltration, soluble salts migrate from the surface to lower soil horizons. Leaching of salts from saturated brine mud (EC >200 mmhos/cm) threatens ground water under the pit (Murphy and Kehew, 1984). However, as previously discussed, EC >200 exceeds recommended soil/waste mixture criteria, Tables A-I-1 and 7-6.

Further, water and associated dissolved constituents move through the soil by natural redistribution governed by pore dynamics, dispersion and diffusion. This process is similar to the method by which extracts move through a long capillary associated with a chromatograph.

Owens et al. (1985) and Bruce et al. (1985) demonstrate this phenomenon. Both studies observed redistribution of surface applied bromide (Br) by rain infiltration and percolation. Owens et al. (1985) showed more than a 7-fold decrease in bromide after percolating water through 2.4 meters (94 inches) of well-drained silt loam. Assuming the same soil conditions, a surface loading of salt equivalent to a soil/salt mixture equal to EC = 8 mmhos/cm (TDS = 4904) results in an EC = 0.7 mmhos/cm at 2.4 m (94 inches). Additionally, assuming sodium chloride salt, the chloride concentration at 2.4 m (94 inches) corresponds to TDS 426 mg/liter. Further, Bruce et al. (1985) demonstrated redistribution of surface bromide (Br) as high as 1800 mg/l resulted in concentrations of <20 mg/liter at 3 meters (118 inches) after nearly 4 years of leaching and 4.7 meters (46 inches/year) of rainfall. Bromide concentration decreased to 100 mg/liter at 1.5 meters and to less than detection limit at 3.8 meters (12 feet). We consider bromide and chloride identically acting materials in soil. Accordingly, at EC - 4 or 8 mmhos/cm, chloride concentrations in the percolating water will be less than the Secondary Drinking Water Quality standard of 250 mg/liter (40 CFR §143.3) at less than 3 feet from the surface source. Thus, the selection of EC = 4 or 8 depends on land use considerations, Tables A-I-1 and 7-6.

SOIL CHEMISTRY

Based on mass action principles of cation exchange in soils, we expect the percentage of sodium on the exchange complex to be a function of the relative concentrations of sodium and competing cations in soil solution. The same principles hold true for E&P waste solids. The degree of sodium saturation in soils and E&P waste solids is determined from three interdependent soil chemical properties. These are:

- Cation Exchange Capacity (CEC)
- Sodium Adsorption Ratio (SAR)
- Exchangeable Sodium Percentage (ESP)

Cation Exchange Capacity (CEC)

Cation Exchange Capacity (CEC) measures the quantity of cations reversibly adsorbed per unit weight of soil. Typically, CEC is expressed in milliequivalents per 100 grams of mass (meq/100 g). The equivalent standard metric unit is centimoles per kilogram (cmole/kg). The predominant cations held on the particle surface of soils and earthen waste solids are calcium (Ca), magnesium (Mg), sodium (Na) and potassium (K), known as the basic cations. The percentage of the CEC occupied by the basic cations is called the base saturation. Fertile soils possess base saturations greater than 80%, consisting predominately of calcium and magnesium cations.

CEC determination requires laboratory analysis (Annex E) of field samples (Annex F). A direct relationship exists between CEC and Exchangeable Sodium Percentage (ESP):

$$ESP = (NaX/CEC) \times 100 \qquad \text{Eq. A-I-3}$$

where NaX (exchangeable sodium) (Annex G) and CEC are expressed in meq/100g. ESP is calculated by an empirical relationship dependent on Sodium Adsorption Ratio (SAR) as shown in the next section.

Sodium Adsorption Ratio

Good soil structure and fertility require large amounts of calcium and magnesium. These beneficial cations form salt complexes of low solubility within soils. On the other hand, sodium cations form soluble salts which dominate soil solutions resulting in deleterious impacts. Accordingly, we require a measure of impact of sodium, i.e., Sodium Adsorption Ratio (SAR).

The U.S. Department of Agriculture Salinity Laboratory (1954) developed SAR (Annex D) as an imperial index of detrimental sodium effects on soils.

$$SAR = Na/[(Ca + Mg)/2]^{1/2} \qquad \text{Eq. A-I-4}$$

where the cation values from direct chemical analysis of waste liquids or aqueous extracts of waste solids or soils are expressed in meq/liter (Annex A). To convert concentrations from mg/liter to meq/liter simply divide the former by the milliequivalent weight of a specific cation (see 4.1 Annex D). The U.S. Salinity Laboratory staff (1954) developed an empirical equilibrium expression relating SAR and ESP.

$$ESP = \frac{100\,(-0.0126 + 0.01475\,SAR)}{[1 + (-0.0126 + 0.01475\,SAR)]} \qquad \text{Equation A-I-5}$$

Exchangeable Sodium Percentage

Knowing the Sodium Adsorption Ratio (Equation A-I-4) the ESP under equilibrium conditions can be determined from Equation A-I-5. Some inaccuracy results when using formula A-I-5 on fresh releases since equilibrium has not been achieved. However, the error is not significant. Accordingly, we recommend using the relationship.

Soil Chemistry Impacts

SAR values greater than 12 indicate high sodium concentrations in soil liquids. This causes Ca and Mg deficiencies in plants because of antagonistic reactions and reduction of solubilities (Kramer, 1969; U.S. Salinity Laboratory staff, 1954).

Sodic describes soil for which ESP>15% (U.S. Salinity Laboratory staff, 1954). Soils exposed to solutions having high salt concentrations so the soil extract SAR >12.8 and ESP >15% risk becoming sodic. The distinguishing feature of sodic soil is lack of structure (tilth and permeability) and tendency for the soil to disperse in water. A dispersed soil devastates plants by reducing air and water supplies to the root system (Reeve and Fireman, 1967; Bresler et al. 1983).

Tucker (1985) demonstrated spent drilling fluid amended soils having SAR <15 and ESP <15% are required for normal plant growth and acceptable soil structure. SAR of soils can be altered easily by soil amendments (gypsum, calcium nitrate, etc.). Accordingly, SAR is less critical than

ESP. Deuel and Brown (1980) showed the detrimental impact on soil when exposed to water having an EC = 2.8 mmhos/cm and SAR = 16.1 was directly proportioned to the calcium in the soil. The occurrence of naturally occurring or amendment excess gypsum in the soil permits disposal of sodic E&P wastes, if the ionic strength of the salt is low. Freeman and Deuel (1984) demonstrated successful pit closures (SAR <15 and ESP <15%) by land incorporation of E&P waste solids having SAR's >200 and ESP's >90 when salt levels in the waste solids after incorporation were <4 EC mmhos/cm. Incorporation consisted of blending waste solids with background soils at defined mixture ratios in the presence of gypsum and fertilizer.

SOIL CHEMISTRY LIMITING VALUES

Based on the above discussion, we recommend limiting the SAR to less than 12 and ESP to less than 15%, Table A-I-1. The Department of Agriculture (U.S. Salinity Laboratory, 1954) and various investigators accept these soil chemistry limits as a means of preventing soil sodicity. However, note the recommended values apply to the waste amended soil and not to the waste per se.

HYDROCARBON IMPACTS

E&P wastes contain crude oil, diesel fuel and lubricating oil (Miller, et al. 1980; Thoresen and Hinds, 1983; Whitfill and Boyd, 1987). Typically, crude oil containing wastes result from well pulling operations, pipeline leaks or overflows. Also, during drilling operations, operators add crude oil or diesel lubricate the bit or decrease water loss from water base muds. Additionally, lubricating oil drips from drilling and producing machinery while in operation.

Hydrocarbon concentrations in water base muds seldom exceed 4% by weight (Freeman and Deuel, 1984). Tank bottoms, emulsions and oil contaminated soil typically contain high concentrations of hydrocarbons.

Crude oil and diesel fuel consist of complex saturated and aromatic hydrocarbons (Thoresen and Hinds, 1983; Oudot et al., 1989). Being a refined product, lubricating oil is a saturated viscous hydrocarbon containing small quantities of additives. Brown, et al. (1983) showed these hydrocarbons easily partition from water by solvent using separatory funnels or extracted from solids using a Soxhlet apparatus. Hydrocarbons extracted

Plate 1 Silica tetrahedron

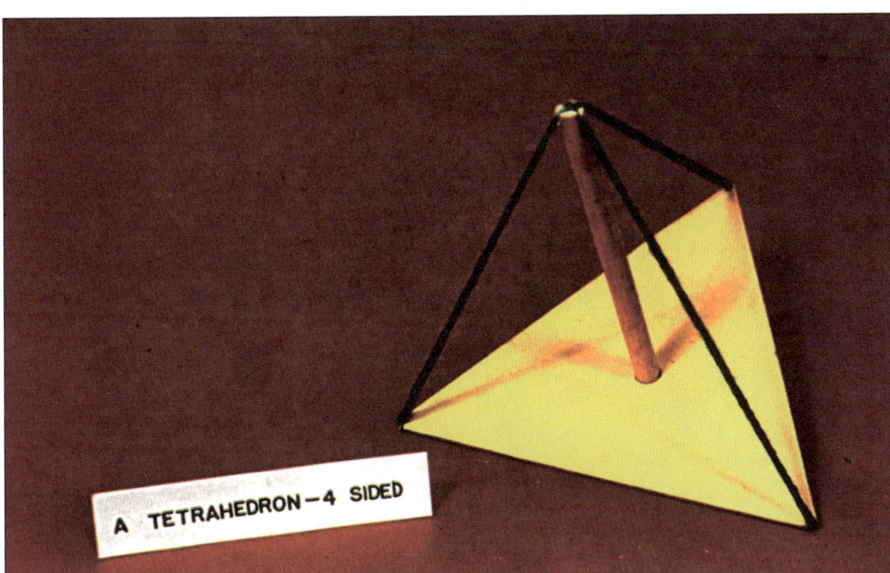

Plate 2 Tetrahedron

Plate 3 Aluminum octahedron

Plate 4 Octahedral arrangement of aluminum hydroxide in kaolinite and gibbsite

Plate 5 Chemical and structural features of kaolinite

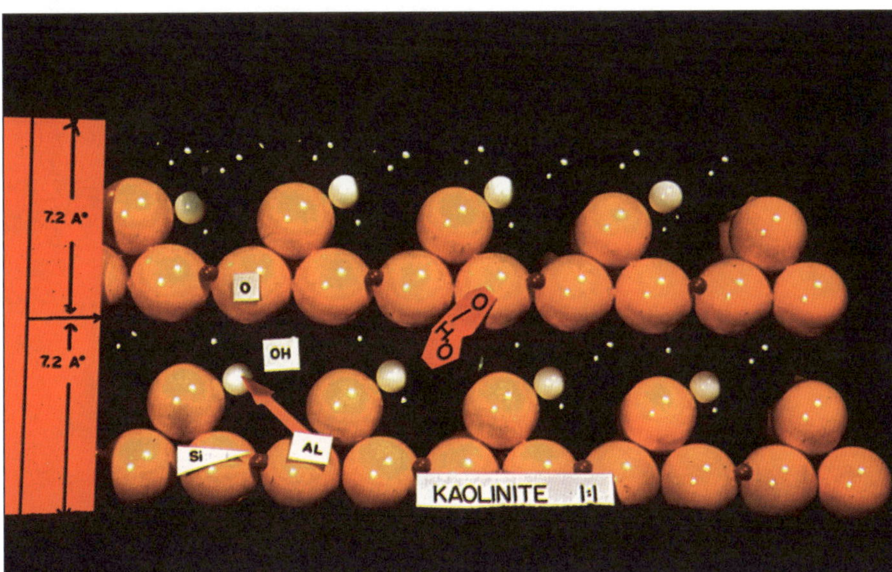

Plate 6 Hydrogen bonding between layers of kaolinite clay

Plate 7 Illite clay

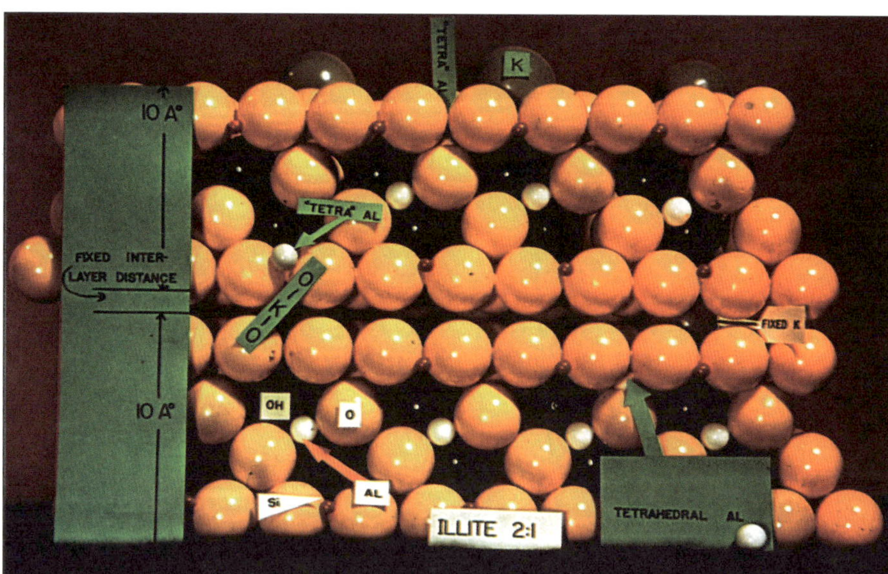

Plate 8 Small fixed interlayer space of illite prevents water and cations from entering

Plate 9 Substitution of magnesium for aluminum in montmorillonite increases CEC

Plate 10 The low attraction between oxygen atoms in adjacent layers of montmorillonite allow free movement of water and exchange of cations

Plate 11 Cations such as magnesium and calcium are hydrated by 6 molecules of water

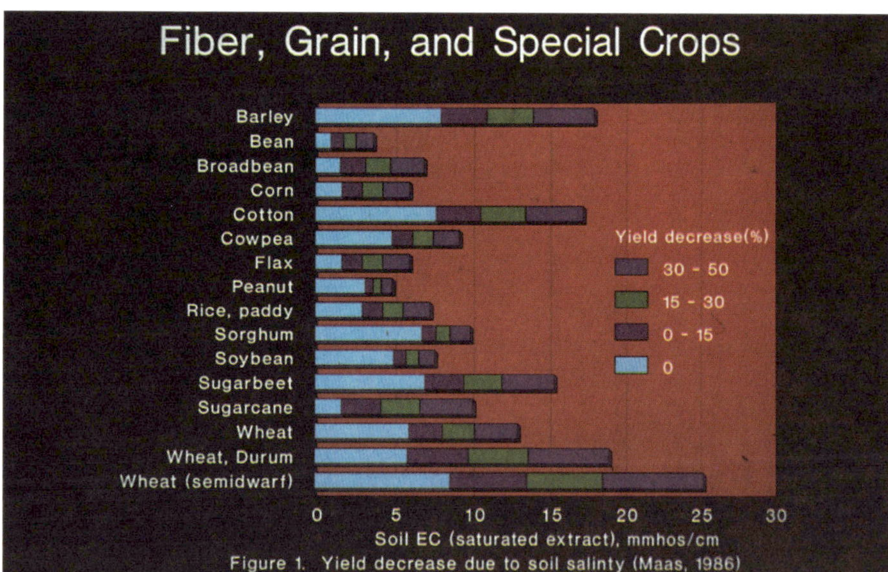

Plate 12 Yield decrease due to soil salinity

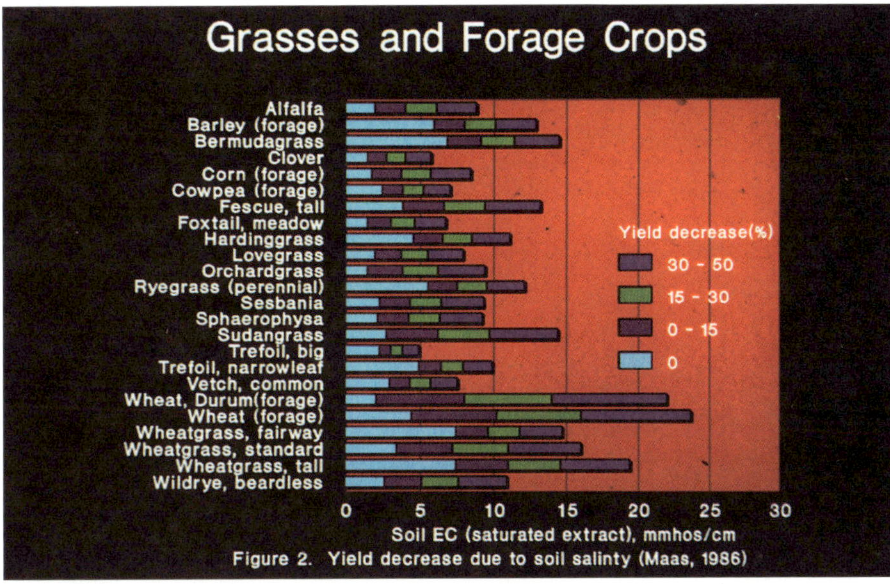

Plate 13 Yield decrease due to soil salinity

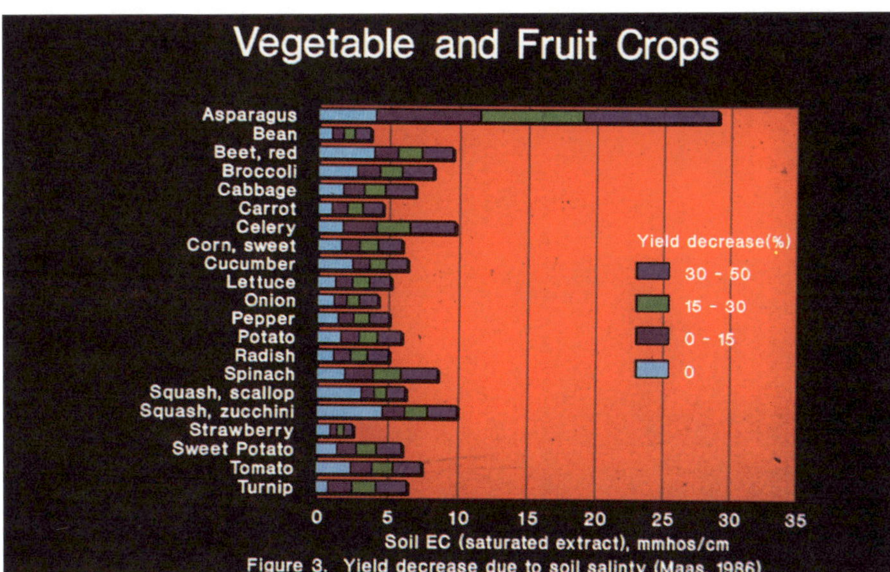

Plate 14 Yield decrease due to soil salinity

Plate 15 Barge pit for Louisiana State water drilling well

Plate 16 Sampling a North Slope above grade (250′w x 300′l x 8′ d) reserve pit serving multiple wells

are assayed gravimetrically and reported collectively as oil and grease. The EPA protocol requires Freon 113 as the extractant, (SW-846, Method 3540). However, for solids, the solvent of choice is methylene chloride because it efficiently extracts petroleum hydrocarbons without co-existing significant quantities of naturally occurring organic matter (Brown and Deuel, 1983).

As discussed before, Total Petroleum Hydrocarbon by Gas Chromatography with Flame Ionization Detector (TPH-GC-FID) is an excellent substitute for Oil and Grease. Typically, we use TPH in place of Oil and Grease especially when demonstrating bioremediation.

Extensive research has occurred regarding the impact of crude oil and diesel fuel on plants and soils (Schollenberger, 1968; Harper, 1939; Plice 1948; Schevendinger, 1968; Garner, 1971; Odu, 1972; Miller and Honarvar, 1975; and Miller, et al. 1980).

Phytotoxic compounds consist of low molecular weight aromatic hydrocarbons present initially (diesel fuel) or form as metabolites of various degradation processes. (Plice (1948), Baker, (1970), Patrick (1971), Honarvar (1975), Udo and Fayemi (1975), and Thoresen and Hinds, (1983) report germination inhibition and yield reduction of row crops planted in soil containing more than 2% crude oil by weight. Yields decreased 50% from control area yields at 2% oil. Pal and Overcash (1978) reported reduction in growth of vegetables and row crops with oil contaminated soil containing 1% by weight. Bulman and Scroggins (1988) demonstrated satisfactory plant growth at 3.5% oil by weight but poor plant growth at oil contents exceeding 5%. They observed, at another site, reduced plant growth during the first season in soil contaminated with 1 and 2%. However, they report enhanced crop growth at 0.5% oil. Frankenberger and Johanson (1982) reported heavy ends of crude oil and refined products added to soil at 20 to 60% by weight disrupt oxidative and soil microflora activities requisite to biological remediation. These observations followed oil spills to soil.

Miller et al. (1980) observed 1% by weight soil loading with diesel fuel resulted in 49 to 69% decrease in yield of beans and corn, respectively. However, replanting after 4 months resulted in normal growth. Younkin and Johnson (1980) planted reed canary grass in soil contaminated with 0.45% diesel fuel. The resulting germination decreased 69% and yield decreased 79%. During the harvest, Younkin and Johnson (1985) observed no decrease in yield (75 days after addition of diesel to the soil). Overcash and Pal (1979) determined 1% oil by weight as the threshold for reduced crop yield. Concentrations of 1.5 to 2% by weight reduced yield by greater than 50%. These effects occur immediately after oil application and before the oil is biodegraded or evaporated. Further, Overcash and Pal (1979) showed tolerance to oil is a function of crop species, Table A-I-3.

Table A-I-3 Oil tolerance for selected crops

Crop Type	Single Oil Application
yams, carrots, rape, lawngrasses, sugarbeets	< 0.5% of soil weight
ryegrass, oat, barley, corn, wheat, beans, soybeans, tomato	< 1.5% of soil weight
red clover, peas, cotton, potato, sorghum	< 3.0% of soil weight
perennial grasses, coastal bermudagrass, trees, plantain	> 3.0% of soil weight

These studies indicate that under hydrocarbon loadings > 1% oil & grease, E&P wastes may be detrimental toward plant growth. However, at 1% or less of mixed hydrocarbons, little or no yield reduction is expected based on existing information. Also, recovery of a site adversely impacted at a 1% loading rate is expected after a few months but not exceeding one growing season, following a one-time application.

Carr (1919) suggested oil damage resulted from reduced aeration-water interaction rather than oil induced toxicity. Ellis and Adams (1961) suggest iron and manganese release under low oxygen conditions contributes to phytotoxicity of soil contaminated with petroleum hydrocarbons. Phytotoxicity appear to reduce after assimilation of the oil by the soil.

We conclude from these studies, E&P wastes mixed at 1% or less with soil presents little or no adverse effect plant yield. Further, recovery of a site contaminated at a loading rate of more than 1% occurs within a few months of the oil application.

Oil Mobility

Several investigators report low mobility of hydrocarbons in soil, particularly in clay soils. Plice (1948) observed natural segregation when liquid oil enters soil. The heavy ends (more viscous fraction) remain near the surface while the light ends penetrate more deeply. This means the more soluble fractions increase with depth (Duffey et al., 1977; Weldon (1978).

Benzene, toluene, ethylbenzene and xylenes (BTEX) represent the more mobile hydrocarbons in soil (Roy and Griffin, 1985). EPA (1987) showed produced water contain significant levels of BTEX, Diesel oil base muds contain BETX but at concentrations low enough to assure attenuation by adsorption onto soil particles (El-Dib et al., 1978). Hydrocarbon mobility is restricted, further, by chromatographic effect of liquid moving through a porous media (Warden, Groenewould and Bridie, 1977). Further, oil floats on water. Accordingly, oil movement through soil is restricted to pores of sufficient diameter, not water saturated. Additionally, the "Jamin effect" (obstruction of a non-wetting fluid flow in porous media) restricts hydrocarbon movement (Schiegg, 1980).

Leaching of hydrocarbons from low concentration contaminated surface soil presents no problem. Watts et al. (1982) observed no oil migration at 30 to 45 cm (12 to 18 inches) depth when 14% oily industrial waste was incorporated in the first 15 cm (6 inches) of surface soil. Raymond et al. (1976) found 99% of the oil remained within the near surface 20 cm. (8 inches) after 1 year when 2% oil was added to the top 15 cm (6 inches) of soil. Streebin et al. (1985) observed no significant oil migration below the zone of incorporation with loading rates of 3 and 13% by weight. Oudot (1989) showed only slight tendency of hydrocarbon leaching at a loading of 2% in soil. Also, hydrocarbon leaching is reduced in areas of high clay soils. Accordingly, the one-time oil concentrations presented in Tables A-I-1and 7-6 represent defensible limits.

Bioremediation

Raymond et al. (1967), Atlas and Bartha (1972), Jobson et al. (1972, Kincannon (1972, Westlake et al. (1974), Horowitz et al. (1975) and Sveinung et al. (1986) observed soils naturally contain adequate diversity of microbes and capacity to biodegrade saturates and light end aromatics (the more mobile fraction). Concentration and composition of hydrocarbons, nutrients, oxygen, moisture and temperature control the rate of degradation (Schwendinger, 1968; Francke and Clark, 1974; Huddleston and Myers, 1978; Dibble and Bartha, 1979; Brown et al., 1983; Flowers et al., 1984; Bleckmann et al., 1989). Specifically, the highest rate of bioremediation occurs when:

- carbon/nitrogen ratio = 1:60-100
- moisture content; 60-80% retained by soil at 0.33 bar pressure, and
- temperature = 35-38° C (95-100° F.)

Watts et al. (1982) measured a 2 year half life for hydrocarbons from a soil contaminated with 14% by weight oil. Streebin et al., (1985) observed a half life of about 2 years at similar loading rate. Oudot et al. (1989) observed 99% field biodegradation of the hydrocarbons in soil loaded at 2% after 3.5 years. Lynch and Genes (1987) measured a half life of 77 days in field plots contaminated with 5% polyaromatic hydrocarbons.

Degradation processes attenuate the mobile, light end aromatics and water soluble petroleum hydrocarbons when applies to the soil surface (Raymond, 1975; Brown et al., 1983; Brown and Deuel, 1983; Whitfill and

Boyd, 1987; Bleckmann et al., 1989). This demonstrates little potential for contaminant migration. Several studies (Plice, 1948; Mackin, 1950; Ellis and Adams, 1961; Baker, 1970; Giddens, 1976) demonstrate low concentration oil applications improve soil physical condition and fertility because the degradation product of oil is soil humus.

The limiting threshold criteria depends on land use and method of disposal. These properties are incorporated in the concepts used to develop Tables A-I-1 and 7-6.

RESERVE PIT OPERATION

Drilling fluids function because of chemical additives which deflocculate (disperse) clay particles. This same property results in forming a filter cake on the bore hole surface and the reserve pit surface upon discharge of spent drilling fluids to the pit.

Pit Surface Sealing

Rowsell et al. (1985) observed dispersed particulate material in manure waste formed an effective seal against contaminant migration in unlined, earthen storage impoundments. The mechanism of sealing was a physical blocking of pores by the particulate matter. This same dispersion and physical blockage occurs when drilling fluids are discharged to the reserve pit. Additionally, clay and fine silt incorporated in the mud during formulation or drilling are incorporated within the filter cake formed when excess mud is discharged with the cuttings to the pit. These particles form a natural liner on the pit surface. The "natural liner" forms a physical barrier having chemisorptive properties. Recent unpublished data support the premise that heavy metals, organics and even chlorides do not permeate the "natural liner".

Reserve pits constructed in course textured soils require more time to form a natural liner. Coating the pit surface with bentonite clay provides an initial "seed" surface for forming the natural liner. Additionally, in arid areas containing loamy or clay soils, a natural liner is slower to form. This allows deeper penetration of pit waste liquids. Prewetting the pit surface after coating with bentonite in sandy soils, or in loamy and clay soils in arid areas, speeds the formation of a seal, and reduces the depth of penetration and fluid loss.

Summary of Thresholds and Applications

Table A-1-4 provides a summary of criteria and applications based on waste type.

Table A-1-4 Summary of E&P waste, disposal technique, and operative criteria

E & P Waste	Disposal Technique	Criteria			
		EC mmho/cm	SAR	ESP %	O&G %
Liquid	roadspreading	4	NA*	NA*	NA*
	landspreading	4	12	15	1
Solids	land treating	4	12	15	1
	burial or landfill	4	NA*	NA*	1

NA* - not applicable

E&P pit wastes consist of liquids and solid phases. Solids have utility as construction fill in arid and semi-arid areas. However, pit solids do not make suitable weight bearing or road surfaces without amendments.

Accordingly, roadspreading is not recommended for these solids.

FLOW DIAGRAMS

Pit Liquids

Pit liquid disposal options depend on the analysis of waste, Figure A-I-1.

Pit Solids

Pit solids evaluation is more complex than required for pit liquids, Figure A-I-2.

SAMPLE CALCULATIONS

Parameters and Example Calculations for Management of Pit Wastes

The following example demonstrates the calculation method for managing reserve pit wastes land treatment. The pit contains liquid and solids. Also, sufficient native soil is available for land treatment. Sampling and analysis provides pit and native soil characteristics, Table A-I-5.

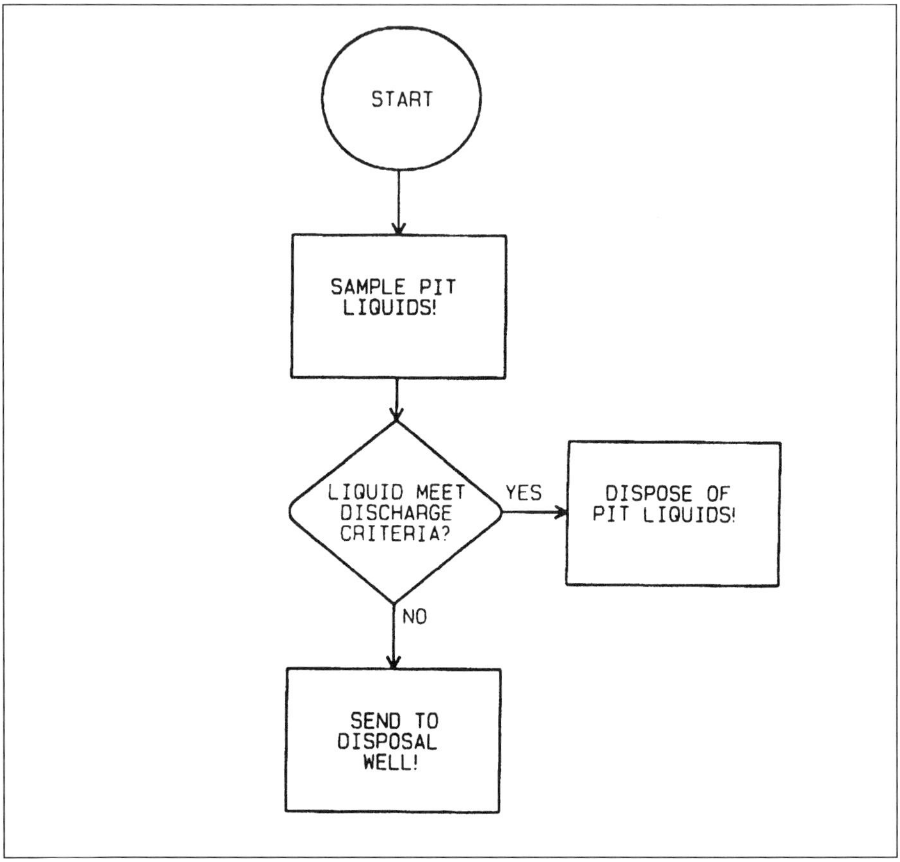

Fig. A-I-1 Steps for pit liquid evaluation and disposal

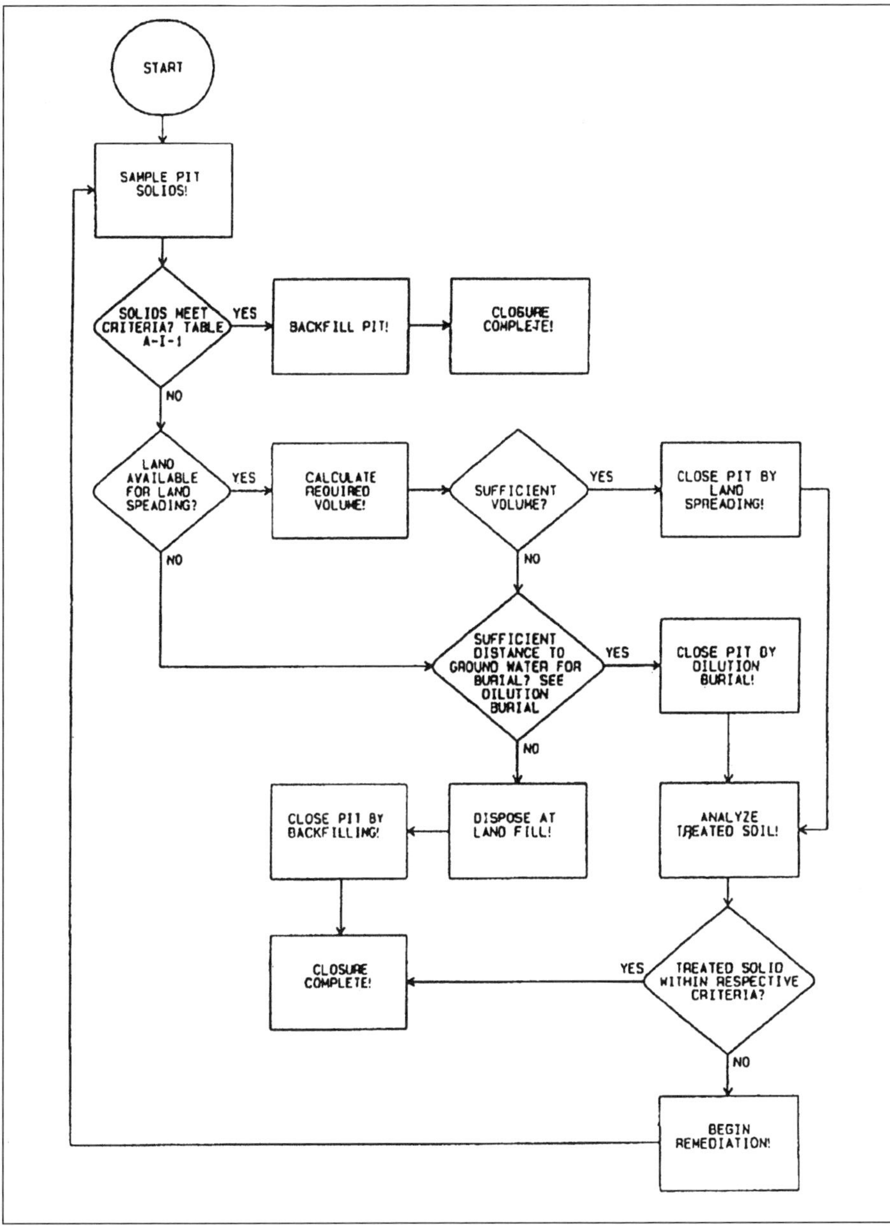

Fig. A-I-2 Steps for pit solids evaluation and disposal

Table A-I-5 Pit material and native soil characteristics

Parameter*	**Pit Liquid****	**Pit Solids†**	**Native Soil†**	**Threshold Level††**
Moisture,%	N/A	243.0	N/A	N/A
TDS, mg/liter	1,410	24,830	272	N/A
EC, mmhos/cm	2.3	40.5	0.4	4
SAR, ratio	4	25	<1	12
Na, meq/liter	9.3	260.0	0.4	N/A
Ca, meq/liter	10.9	199.0	2.5	N/A
Mg, meq/liter	1.7	16.0	0.9	N/A
CEC, meq/100g	N/A	13.5	39.6	N/A
Na, meq/100 g	N/A	2.8	0.3	N/A
Ca, meq/100 g	N/A	18.5	24.8	N/A
Mg, meq/100 g	N/A	0.3	7.5	N/A
ESP,%	N/A	20.7	<0.1	12
O&G,%	0.2	10.1	<0.1	1
Volume, bbl	12,938	21,897	N/A	N/A

NA = not applicable

* Parameters are reported on a dry weight basis unless noted otherwise.

** NA - means the parameter meets the guidance threshold or is not applicable for that matrix.

† Soluble constituents were determined for saturated paste extracts of pit solids and native soil.

†† An ESP of 12% is recommended in establishing land requirements for Na management.

Determination of Limiting Constituent(s)

The following determines the limiting constituents for pit solids and pit liquids management:

Pit Liquid Management

a) Comparison of pit liquid analyses and threshold values show no chemical limitation for land application.

b) Native soil loading capacity for Na using an ESP of 12% and materials distribution depth of 6 in/acre.

Given : 1 acre-6 in = 2,000,000 lb.

1 mg/kg = 1 lb. / 1,000,000 lb.

measured CEC = 39.6 meq/100g, Table A-I-5
allowable ESP = 12% = 0.12
Na = 23 mg/meq (Annex D)

Then

Na, mg/kg = CEC × ESP × Na
= (39.6 meq/100g) × 0.12 × (23 mg/meq) ×(1000 g/kg)
= 39.6 × 0.12 × 23 × 10
= 1093 mg/kg
Since 1 mg/kg = 1 lb/1,000,000 lb.
and 1 acre - 6 in = 2,000,000 lbs

Then

Na, lb/acre-6 in = $1093 \times 2 \times 10^6 / 1 \times 10^6$
= 2185 lb/acre-6 in

c) Mass Na in pit liquids
Na = 9.3 meq/l
pit volume = 12,938 bbl
Na =

$$\frac{[(9.3\text{meq/l}) \times (23\text{mg/meq})\text{x}(3.8\text{l/gal}) \times (42\text{gal/bbl}) \times (12938\text{bbl/pit})]}{[(1000\text{mg/g}) \times (454\text{g/lb})]}$$

= 973

d) Land area required sodium distribution, assuming 6 inch depth.
Area = (973 lb Na)/(2186 lb Na/acre-6 in)
= 0.45 acre - 6 inches

e) Liquid management limitation
12,938 bbl of liquid

$$\text{Area} = \frac{(12{,}938\text{bbl}) \times (42\text{gal/bbl})\text{acre ft}^3}{43560\text{ft}^2 \times \dfrac{1\text{ ft}}{12\text{in.}} \times 7.48\text{gal}}$$

= 20 acre-in

f) Assuming the soil infiltration rate = 1.12 in/hr but drops to less than 0.1 in/hr within 10 minutes. This means that once the surface becomes wet, infiltration is limited by the movement of water or drainage within the profile. The dry surface can receive about 0.3 inches without producing a runoff. The land area required for the pit liquids is:

Area = 20 acre-in/0.3 in
= 67 acres

Constructing a ring levee allows infiltration of liquid at a greater depth over a longer time while using a smaller area.

Pit Solids Management

a) Compare EC, SAR, ESP and Oil and Grease for pit solids and limiting values. This shows EC, ESP and Oil and Grease are potential limiting constituents. The high calcium content of the solids and receiving soil suggests SAR is not limiting.

b) Percent solids by weight solids analysis shows solids contain 243% moisture.

$$\text{Dry weight, g} = \frac{(100\text{g wet wt}) \times 100}{(100{+}243\%\text{ moisture})}$$

$$= 29.15\text{ g}$$

$$\text{Solids,\%} = \frac{29.5\text{g}}{100\text{g}} \times (100) = 29.15\%$$

c) Solids volume

$$\text{Volume, bbl} = 21{,}897\text{ bbl wet} \times \frac{29.15}{100}$$

$$= 6383\text{ bbl}$$

d) Total Dissolved Solids

Allowable EC = 4 for land application

TDS = $4 \times 613 = 2452$ mg/l

Using proportionately rule (chapter 1), the allowable TDS (2452 mg/l) results from combining native soil and pit solids, on a dry weight basis.

$$2{,}452\text{ mg/l} = \frac{(6{,}383\text{ bbl})(24{,}830\text{ mg/l}) + (X\text{ bbl})(272\text{mg/l})}{(6{,}383\text{ bbl} + X\text{ bbl})}$$

Solving for X = Volume of native soil

(6383 bbl dry solids) + (X bbl native soil) $\times$ (2452 mg/l) =
(6383 bbl dry solids) $\times$ (24,830 mg/l) + (X bbl native soil) x (272 mg/l)

Volume of native soils = 65,522 bbl

$$\text{V, acre-6 in} = \frac{65{,}532\text{ bbl} \times 5.614\text{ cuft/bbl}}{43560\text{ sqft/acre} \times \frac{\text{ft}}{2(6\text{in})}}$$

Volume = 16.89 acre - 6 in

Thus 16.9 acres are required to reduce the TDS to the acceptable limit.

e) EC Land Requirements
Allowable EC = 4 mmhos/cm for land application.

$$4\text{ EC} = \frac{(40.5\text{ EC})(6{,}383\text{ bbl}) + (0.4\text{ EC})(X\text{ bbl})}{(6{,}383\text{ bbl} + X\text{ bbl})}$$

X = 64,717 bbl native soil

$$V,\text{ acre - 6 in} = \frac{64{,}717\text{ bbl} \times 5.614\text{ cuft/bbl}}{43560\text{ sqft/acre} \times \frac{\text{ft}}{2(6\text{in})}} = 16.68\text{ acre-6 in}$$

Accordingly, 16.7 acre-6 in are required to reduce EC to the acceptable limit.

f) ESP Land Requirement
Allowable ESP = 12% Pit solids ESP = 20.7%

$$12\% = \frac{(20.7\%)\,(6{,}383\text{ bbl}) + (x)\,(0.1\%)}{(6{,}383\text{ bbl}) + X}$$

$$X = \frac{4{,}667\text{ bbl} \times 5.61\text{ ft}^3/\text{bbl}}{(43{,}560\text{ ft2/acre} \quad \frac{\times\text{ ft}}{2/6\text{ in.}})}$$

$$= 1.2\text{ acre -6 in.}$$

Accordingly, only 1.2 acre - 6 in are required for land treatment of ESP.

g) Oil and Grease Land Requirement
Allowable oil and grease = 1%
Pit solids oil and grease = 10.1%
Native soil oil and grease = 0.01%

$$1\%\text{ O\&G} = \frac{(10.1\%\text{ O\&G})(6{,}383\text{ bbl}) + (0.01\%\text{ O\&G})(X\text{ bbl})}{(6{,}383\text{ bbl} + X\text{ bbl})}$$

X = 58,672 bbl
= 15.1 acre - 6 in

Accordingly, 15.1 acre - 6 in are required for land treatment of oil and grease.

h) Limiting Criteria
Comparing the land requirement for EC, ESP and Oil & Grease, it is clear that EC requires the most acreage, 16.89 acre - 6 In Accordingly, EC is the limiting constituent. Spreading the pit solids over an area consisting of 16.89 acres, allowing the solids to dry and disking and cross-disking to 6 in satisfies the criteria.

i) Landspreading Solids
Placing wet solids too deep increases the probability of equipment becoming immobilized. Accordingly, spread solids no more than 4 inches thick.

$$\text{Thickness of wet solids} = \frac{(21897 \text{ bbl wet solids(acre–ft)}}{(16.89 \text{ acre}) \times (43560 \text{ cuft})} \times \frac{(5.614 \text{ cuft})12\text{i}}{(\text{bbl})\text{ft}}$$

$= 2.0$ in

Accordingly, spread wet solids to about 2 inches thick and let dry. Then disk and cross-disk.

j) Nitrogen Addition
The addition of nutrients to the admixed soil encourages plant growth. Add nitrogen in proportion to the carbon content of the admixed soil. As discussed in Chapter 9, oil contains about 78% carbon. Phosphorous (P) and Potassium (K) are added at a ratio of N:P:K = 4:1:1. The carbon to nitrogen ratio = 150:1.

Oil and grease in the admixture = 1% oil and grease weight/acre
= (1% oil and grease) x (10000 ppm/%) x 2 ppm/lb/ac)
= 20,000 lb/acre

$$\begin{aligned} \text{Carbon weight/acre} &= (20{,}000 \text{ lb/acre}) \times 0.78 \\ &= 15{,}600 \text{ lb carbon/acre} \\ \text{N requirement/acre} &= 15{,}600/150 \\ &= 104 \text{ lb/acre} \end{aligned}$$

$$\text{For urea (45\% N) fertilizer required} = \frac{104}{0.45} = 231 \text{ lb/acre}$$

$$\text{For ammonium nitrate (33\%) fertilizer} = \frac{104}{0.33} = 315 \text{ lb/acre}$$

References

API (American Petroleum Institute). 1987. "Oil and Gas Industry Exploration and Production Wastes." DOC No. 471-01-09, July.

Atlas, R. M., and R. Bartha, 1972. "Degradation and Mineralization of Petroleum by Two Bacteria Isolated From Coastal Waters." *Biotechnol. & Bioeng.* 14:297-308.

Baker, J. M. 1970. "The Effects of Oil on Plants." *Environ. Pollution.* 1:27-66.

Baldwin, I. L. 1922. "Modifications of Soil Flora Induced by Application of Crude Petroleum." *Soil Sci.* 14:465-477.

Barrow, G. M. 1966. "The Nature of Electrolytes in Solution.". *Physical Chemistry.* McGgraw-Hill Book Co., New York. 648-684

Bleckmann, C. A., L. J. Gawel, D. L. Whitfill, and C. M. Swindoll. 1989. "Land Treatment of Oil-Based Drill Cuttings." SPE/IADC Drilling Conf. Proc. New Orleans, LA. 529-536.

Bresler, E., B. L. McNeal, and D. L. Carter. 1983. *Saline and Sodic Soils.* Springer-Verlag Publishing Co. Berlin, Heidelberg, and New York.

Brown, K. W., L. E. Deuel, Jr. 1983. "An Evaluation of Subsurface Conditions at Refinery Land Treatment Sites." EPA-600/52-83-096.

Brown, K. W., L. E. Deuel, Jr., and J. C. Thomas. 1983. "Land Treatability of Refinery and Petrochemical Sludges." EPA-600/52-83-074.

Bruce, R. R., R. A. Leonard, A. W. Thomas, and W. A. Jackson. 1985. "Redistribution of Bromide by Rainfall Infiltration into a Cecil Sandy Loam Landscape." *J. Environ. Qual.* 14:439-445.

Bulman, T. L., and R. P. Scroggins. 1988. "Environment Canada Research on Land Treatment of Petroleum Wastes." 81st Annual Meeting, APCA. Dallas, Texas.

Carr, R. H., 1919. "Vegetable Growth in Soil Containing Crude Petroleum." *Soil Sci.* 8:67-68.

Deuel, L. E., Jr. and K. W. Brown. 1980. "The Feasibility of Utilizing Uranium Leach Water as an Irrigation Source. Final Report." Texas A&M University Research Foundation, College Station, Texas.

Deuel L.E., Jr. and G.H. Holliday. 1993. "Determination of Total Petroleum Hydrocarbons in Soil." SPE 26394. October, Houston, TX.

Dibble, J. T., and R. Bartha, 1979. "Effect of Environmental Parameters on Biodegradation of Oil Sludge in Soil." *Soil Sci.* 127:365-370.

Duffy, J. J., M. F. Mohtadi, and E. Peake. 1977. "Subsurface Persistence of Crude Oil Spilled on Land and Its Transport in Groundwater.". J.O. Ludwigson (ed.). Proc. 1977 Oil Spill Conf. New Orleans, LA. API Washington, D.C. p. 475-478

El-Dib, M. A., A. S. Moursy and M. I. Badawy. 1978. "Role of Adsorbents in the Removal of Soluble Aromatic Hydrocarbons from Drinking Waters." Wat. Res. 12:1131-1137.

Ellis, R. E., Jr., and R. S. Adams, 1961. "Contamination of Soils by Petroleum Hydrocarbons." *Advances in Agron.* 13:197-216.

EPA, U. S. Environmental Protection Agency. 1987. "Exploration, Development, and Production of Crude Oil and Natural Gas, Field Sampling and Analysis Results." EPA 530-SW-87-005, Dec.

Ferrante, J. G., 1981. "Fate and Effects of Whole Drilling Fluids and Fluid Components in Terrestrial and Fresh Water Ecosystems." EPA 600/4-81-031.

Flowers, T. H., I. D. Pulford, and H. J. Ducan. 1984. "Studies on the Breakdown of Oil in Soil." *Environ. Pollut.* 8:71-82.

Foth, H. D., and L. M. Turk, 1972. *Fundamentals of Soil Science, 5th Edition.* John Wiley & Sons, Inc., New York, NY.

Frankenberger, W. T., Jr., and C. B. Johanson. 1982. "Influence of Crude Oil and Refined Petroleum Products on Soil Dehydrogenase Activity." *J. Environ. Qual.* 11:602-607.

Francke, H. C., and F. E. Clark, 1974. "Disposal of Oily Waste by Microbial Assimilation." *U.S. Atomic Energy Report.* Y-1934.

Freeman, B. D., and L. E. Deuel, Jr., 1984. "Guidelines for Closing Drilling Waste Fluid Pits in Wetland and Upland Areas." 2nd Industrial Pollution Control Proceedings. 7th ETCE Conference, New Orleans, LA.

Freeman, B. D. and L. E. Deuel, Jr. 1986. "Closure of Freshwater Base Drilling Mud Pits in Wetland and Upland Areas." Proceedings of a National Conference on Drilling Muds. Univ. of Okla. Environ. and Groundwater Institute. Norman, OK. May 29-30. 303-327.

Garner, J. H. 1971. "Changes in Soil and Death of Woody Ornamentals Associated with Heating Natural Gas." *Phytopathology* 61:892.

Giddens, J. 1976. "Spent Motor Oil Effects on Soil and Crops." *J. Environ. Qual.* 5:179-181.

Harper, J. T. 1939. "The Effect of Natural Gas on the Growth of Microorganisms and Accumulation of Nitrogen and Organic Matter in the Soil." *Soil Sci.* 48:461-466.

Haywood, H. E. and C. H. Wadleigh. 1949. "Plant Growth on Saline and Alkali Soils." *Advan. in Agron.* 1:1-38. Agron. Soc. of Amer. Madison, WI.

Hem, J. D. 1985. "Study and Interpretation of Chemical Characteristics of Natural Water." 3rd Edition. U.S. Geological Survey Water Supply Paper 2254. Alexandria, VA.

Henderson, G. 1982. Analysis of Hydrologic and Environmental Effects of Drilling Mud Pits and Produced Water Impoundments. Dames and Moore, Houston, Texas. Report available from API, Dallas, TX.

Horowitz, A., D. Gutrick, and E. Rosenberg, 1975. Sequential Growth of Bacteria on Crude Oil. *Appl. Microbiol.* 30:10-19.

Huddleston, R. L., and J. D. Myers, 1978. "Treatment of Refinery Oily Wastes by Land Farming." 85th Natl. Meeting of Amer. Inst. Chem. Eng. Philadelphia, PA. June 4-8, 1978. Amer. Inst. Chem. Eng., New York, NY.

Jobson, A., F. D. Cook, and W. S. Westlake. 1972. "Microbial Utilization of Crude Oil." *Appl. Microbio*l. 24:1082-1089.

Kincannon, C. B. 1972. "Oily Waste Disposal by Cultivation Process. Office of Research and Monitoring." Rep. No. EPA-R2-72-110. Washington, D.C., 20460.

Kramer, P. J. 1969. *Plant and Soil Water Relationships: A Modern Synthesis.* McGraw-Hill Inc., New York.

Lunin, J. 1967. "Water for Supplemental Irrigation." WaterIn Quality Criteria, ASTM STP 416. 1st National Meeting on Water Quality Criteria, Philadelphia, PA. September. 63-78

Lynch, J.< and B. R. Genes. 1987. "Land Treatment of Hydrocarbon Contaminated Soils." PIn. T. Kostecki and E. J. Calabrese (eds.) *Petroleum Contaminated Soils.* Volume I. Remediation techniques environmental fate risk assessment. Lewis Publishers, Chelsea, MI. 163-174

Maas, E. V., and G. J. Hoffman, 1977. "Crop Salt Tolerance - Current Assessment." *Journal of the Irrigation and Drainage Division Proceedings*, Amer. Soc. of Civil Eng. Vol 103 no. IRZ.

Maas, E. V. 1986. "Salt Tolerance of Plants." *Appl. Agric. Res.* 1:12-26.

Mackin, J. G. 1950. Report on a Study of the Effects of Applications of Crude Petroleum on Salt Grass, Distichlis spicota. L. Green, Texas A&M Res. Foundation Project 9, 8- 54.

Miller, R. W. and S. Honarvar. 1975. "Effect of Drilling Fluid Component Mixtures on Plants and Soil." Environmental Aspects of Chemical Use in Well Drilling Operation, Conference Proceedings, May, Houston, TX. EPA 560/1-75-004, 125-143.

Miller, R. W., S. Honarvar, and B. Hunsaker. 1980. "Effects of Drilling Fluids on Soils and Plants: I. Individual Fluid Components." *J. Environ. Qual.* 9:547-552.

Miller, R. W., and P. Pesaran. 1980. "Effects of Drilling Fluids on Soil and Plants: II. Complete Drilling Fluid Mixtures." *J. Environ. Qual.* 9:552-556.

Murphy, H. F. 1929. "Some Effects of Crude Petroleum on Nitrate Production, Seed Germination and Growth." *Soil Sci.* 27:117- 120.

Murphy, E. C., and A. E. Kehew. 1984. "The Effect of Oil and Gas Well Drilling Fluids on Shallow Groundwater in Western North Dakota." Report No. 82. North Dakota Geological Survey.

Nelson, M., S. Liu, and L. Sommers, 1984. "Extractability and Plant Uptake of Trace Elements from Drilling Fluids." *J. Environ. Qual.* 13:562-566.

Odu, C. T. I., 1972. "Microbiology of Soils Contaminated with Petroleum Hydrocarbons. Extent of Contamination and Some Soil and Microbiological Properties After Contamination." *J. Inst. Petrol.* 58:210-208.

Overcash, M. R., and D. Pal. 1979. "Design of Land Treatment Systems for Industrial Waste - Theory and Practice." *Ann Arbor Sci.*, Ann Arbor, MI.

Oster, J. D., and J. D. Rhoades. 1984. "Water Management for Salinity and Sodicity Control" In*Irrigation with Reclaimed Municipal Wastewater*. Lewis Publishers, Chelsea, MI. Chapter 7, 1-20.

Oudot, J., A. Ambles, S. Bourgeois, C. Gatellier, and N. Sebyera. 1989. "Hydrocarbon Infiltration and Biodegradation in Land Farming Experiment." *Environ. Pollut.* 59:17-40.

Owens, L. B., R. W. Van Kuren, and M. M. Edwards. 1985. "Groundwater Quality Changes Resulting from a Surface Bromide Application to a Pasture." *J. Environ. Qual.* 14:543- 548.

Pal, D., and M. R. Overcash. 1978. "Plant-Oil Assimilation Capacity for Oils." 85th Natl. Meeting of Amer. Inst. Chem. Eng. Philadelphia, PA.

Patrick, Z. A. 1971. "Phytotoxic Substances Associated with the Decomposition in Soil of Plant Residues." *Soil Sci.* 111:13-18.

Plice, M. J., 1948. "Some Effects of Crude Petroleum on Soil Fertility." *Soil Sci.* Soc. Amer. Proc. 13:413-416.

Raymond, R. L., V. W. Jamison and J. O. Hudson, 1967. "Microbial Hydrocarbon Co-oxidation of Mono and Dicyclic Hydrocarbons by Soil Isolates of the Genus Norcardia." *Appl. Microbiol.* 15:357-865.

Raymond, R. L., J. O. Hudson, and V. W. Jamison, 1975. "Assimilation of Oil by Soil Bacteria." Proc. 40th Annual Midyear Meeting, API Refining.

Raymond, R. L., J. O. Hudson, and V. W. Jamison. 1976. "Oil Degradation in Soil." *Applied and Environ. Microbiol.* 31:522-535.

Reeve, R. C., and M. Fireman. 1967. "Salt Problems in Relation to Irrigation." R.CIn. Dinauer (ed.) Irrigation of Agricultural Lands. *Agron.* Am. Soc. of Agron., Madison, WI. 11:908-1008.

Rowsell, J. G., M. H. Miller, and P. H. Groenevelt. 1985. "Self- Sealing of Earthen Liquid Manure Storage Ponds: II. Rate and Mechanism of Sealing." *J. Environ. Qual.* 14:539-543.

Roy, W. R., and R. A. Griffin 1985. "Mobility of Organic Solvents in Water-Saturated Soil Materials." *Environ. Geol. Wat. Sci.* 7:241-247.

Schiegg, H. O. 1980. "Field Infiltration as a Method for the Disposal of Oil-in-Water Emulsions from the Restoration of Oil-Polluted Aquifers." *Wat. Res.* 14:1011-1016.

Schollenberger, C. J., 1930. "Effect of Leaking Natural Gas upon the Soil." *Soil Sci.* 29:260-266.

Schwendinger, R. B., 1968. "Reclamation of Soil Contaminated with Oil." *J. Inst. Petrol.* 54:182-192.

Streebin, L. E., J. M. Robertson, H. M. Schornick, P. T. Bowen, K. M. Bagawandos, A. Habibafsan, T. G. Sprehe, A. L. Callender, C. J. Carpenter, and V. G. McFarland. 1985. " Land Treatment of Petroleum Refinery Sludges." EPA 600/2-84-193.

Sveinung, S., A. Lode, T. A. Pedersen. 1986. "Biodegradation of Oily Sludge in Norwegian Soils." *Appl. Microbiol. Biotechnol.* 23:297-301.

Thoresen, K. M., and A. A. Hinds. 1983. "A Review of the Environmental Acceptability and the Toxicity of Diesel Oil Substitutes in Drilling Fluid Systems." IADC/SPE Drilling Conf. Proc. New Orleans, LA. 343-352.

Tucker, B. B., 1985. "Soil Application of Drilling Wastes." Proceedings of National Conference on Disposal of Drilling Wastes. Univ. of Okla. Environ. and Groundwater Institute. Norman, OK. May 29-30. 102-112.

Udo, E. J., and A. A. A. Fayemi. 1975. "The Effect of Oil Pollution of Soil on Germination, Growth and Nutrient Uptake of Corn." *J. Environ. Qual.* 6:369-372.

U. S. Salinity Laboratory Staff. 1954. "Diagnosis and Improvement of Saline and Alkali Soils." Agric. Handbk. no. 60, USDA. U.S. Government Printing Office, Washington, DC.

Waarden, M. Van Der, W. M. Groenewoud, A. L. A. M. Bridie. 1977. "Transport of Mineral Oil Components to Groundwater II." Wat. Res. II. 359-365.

Watts, J. R., J. C. Corey, and K. W. McLeod. 1982. "Land Application Studies of Industrial Waste Oils. *Environ. Pollut.*(Series A) 28:165-175.

Weldon, R. A. 1978. "Biodegradation of Oily Sludge by Soil Microorganisms." Suntech Project 48-036. Rep. No. 2, API Washington, D.C.

Westlake, D. W. S., A. Jobson, R. Phillippe, and F. D. Cook, 1974. "Biodegradability and Crude Oil Composition." *Can. J. Microbiol.* 20:915-928.

Whitfill, D. L. and P. A. Boyd. 1987. Soil Farming of Oil Mud Drill Cuttings. SPE/IADC Drilling Conf. Proc. New Orleans, LA. p. 429-438.

Younkin, W. E., and D. L. Johnson. 1980. "The Impact of Waste Drilling Fluids on Soils and Vegetation in Alberta." Symp. Proc. Vol. 1: Research on Environmental Fate and Effects of Drilling Fluids and Cuttings. Lake Buene Vista, FL. Jan. 21-24. 98-112.

U. S. EPA 40 CFR Part 143 - "National Secondary Drinking Water Regulations. Section 143.3 - Secondary maximum contaminant levels." July 1, 1984.

ANNEX A

Saturated Paste Extract

1.0 Scope and Applications

1.1. Saturation percentage is a condition of soil related to field moisture and associated plant response. It is reproducible and approximately equivalent to twice the% moisture at field capacity (0.33 bar) and 4 times the% moisture at permanent wilting (15 bar).This method is used to obtain a saturation extract for the following analyses:

1.1.1. TDS

1.1.2. EC

1.1.3. SAR

2.0. Summary of Method

2.1. Add water to a known amount of sample until point where no more water can be added without forming free water layer.

3.0. Interferences

3.1. Excessive stirring puddles the sample and reconstitute the dispersed condition of most E&P waste solids. Puddled soils represent a gross over estimation of the saturation percentage.

4.0. Apparatus

4.1. Container of 250 ml capacity.

4.2. Buchner funnel, filter paper, vacuum source, and collection vessel.

5.0. Procedure

5.1. Weigh 100 g, dried, ground and sieved solids into 250 ml container.

5.2. Add distilled water to fill pores, stirring gently as needed to achieve saturation. The solid:water mixture is consolidated occasionally by tapping container on workbench.

5.3. At saturation the mixture glistens as it reflects light, flows slightly when the container is tipped.

5.4. Allow paste mixture to stand 1 hr. and check for conditions of paste. Mixture should not stiffen nor should free water form at the surface.

5.5. Add a known quantity of solid sample material if free water forms or more distilled water if mixture stiffens.

5.6. Record the weight of water used to achieve saturation and transfer to the vacuum filter apparatus. Vacuum extraction should be terminated when air begins to pass through the filter.

5.7. Extract is used to measure TDS, EC and SAR

6.0. Calculation

Saturation Percentage (SP)% = $(W - D)/(D - P) \times 100$

where

W = wet weight of sample + pan, g

D = dry weight of sample + pan, g

P = weight of pan, g

7.0. References

7.1. U. S. Salinity Laboratory Staff. 1954. "Diagnosis and Improvement of Saline and Alkali Soils." Agriculture Handbook 60.

ANNEX B

EC

1.0. Scope and Application

1.1. Electrical conductivity indicates the quantity of soluble salts in an aqueous sample. This method applies to pit liquids and saturated paste extracts.

2.0. Summary of Method

2.1. EC is measured directly with the reading corrected to specific conductance at 25°C.

3.0. Apparatus and Materials

3.1. Temperature compensating conductivity meter

3.2. Conductivity cell

3.3. Reagents

3.3.1. ASTM Type II water

3.3.2. 0.01 N potassium chloride

4.0. Procedure

4.1. Rinse conductivity cell and fill with calibration standard. Read and record conductivity.

4.2. Rinse conductivity cell and fill with sample. Read and record conductivity.

5.0. Calculations

5.1. Cell Constant, C
C = (1.413 mmhos/cm) / (ECKCl mmhos/cm)
where: ECKCl = measured conductance, mmhos/cm

5.2. Specific Conductance of Sample
EC = (ECm)(C)
where: measured conductance of sample, mmhos/cm
C = cell constant

6.0 References

6.1. Rhoades, J.D. 1982. Soluble Salts. A.L Page (ed.) *Methods of Soil Analysis*. Part 2 - Chemical and Microbiological Properties. 2nd Edition. (Ed.) ASA Agronomy Monograph 9. 172-173.

ANNEX C

TDS

1.0. Scope and Application

1.1. This procedure applies to E&P aqueous phase samples including produced water, pit liquids and saturated paste extracts.

2.0. Summary of Method

2.1. Total Dissolved Solid is mineral matter passing a standard glass filter, and remaining after drying at 180°C to constant weight.

3.0. Interferences

3.1. The principle interference is from fine clay fractions and organic colloids passing the filter and stable at 180°C.

4.0. Apparatus and Materials

4.1. Evaporating dishes
4.2. Filtration equipment
4.3. 0.45-μm filters
4.4. Drying oven, for operation to 180°C (+/- 2°C)
4.5. Analytical balance, capable to 0.1 mg

5.0. Procedure

5.1. Assemble filtration equipment and insert 0.45 μm- filter.
5.2. Apply vacuum and wash disk with 3, 20-ml volumes of distilled water. Discard washwater.
5.3. Filter measured volume of homogenized sample through filter, wash with 3, 10-ml volumes of distilled water, allowing complete drainage between washings.
5.4. Transfer filtrate to weighed evaporation dish previously cleaned by ignition to 550°C for 1 hr.
5.5. Evaporate water at 180°C to a constant weight. Evaporation dish is cooled in desiccator prior to weighing.

6.0. Calculation

TDS, mg/liter = (A - B) $\times$ 1000/sample volume, ml
where
A = weight of residue + dish, mg
B = weight of dish, mg

7.0. References

7.1 Standard Methods for the Examination of Water and Wastewater. 1985. 16th Edition. APHA. AWWA. WPCF. Method 209 B. Total Dissolved Solids Dried at 180°C.

ANNEX D

SAR

1.0. Scope and Application

1.1. This method applies to most E&P wastes including pit liquids and water extracts of pit solids or waste solid:soil mixtures.

2.0. Summary of Method

2.1. Soluble cations are determined by atomic absorption spectrophotometry or other suitable instrumentation for pit liquids or water extracts of solid phase samples. The sodium adsorption ratio (SAR) is calculated from the cationic distributions.

3.0. Procedure

3.1. Calibrate instrumentation using standards of known concentration.

3.2. Read concentrations of Na, K, Mg and Ca direct for pit liquid samples or aqueous extracts including saturated pastes.

4.0 Calculations

4.1. Conversion to meq/liter

Cl, meq/liter = (Cl mg/liter) / (35.5 mg/meq)

Na, meq/liter = (Na mg/liter) / (23 mg/meq)

K, meq/liter = (K mg/liter) / (39 mg/meq)

Ca, meq/liter = (Ca mg/liter) / (20 mg/meq)

Mg, meq/liter = (Mg mg/liter) / (12 mg/meq)

4.2. SAR

$$\text{SAR} = (\text{Na, meq/l}) / [(\text{Ca, meq/l} + \text{Mg, meq/l})/2]^{1/2}$$

5.0. References

5.1. Rhoades, J.D. 1982. Soluble Salts. A.L Page (ed.) *Methods of Soil Analysis*. Part 2 - Chemical and Microbiological Properties. 2nd. Edition. ASA Agro. Monograph 9.173-174.

ANNEX E

Cation Exchange Capacity

1.0. Scope and Application

1.1. This method applies to most soils and E&P waste, including calcareous and non-calcareous samples.

2.0. Summary of Method

2.1. The sample is saturated with an excess of sodium acetate solution, resulting in an exchange of other cations by sodium. Subsequently, excess sodium is rinsed from the sample followed by quantitative desorption of sodium by ammonium. The concentration of displaced sodium is then determined by atomic absorption, emission spectroscopy, or an equivalent means as available and approved by EPA.

3.0. Interferences

3.1. Soluble salts and gypsum will interfere with the CEC determination if they are present in sufficient quantities. These may be overcome by washing the solids with water before saturating with sodium, or employ a more exhaustive saturation procedure.

4.0. Apparatus and Materials

4.1. Centrifuge and centrifuge tubes
4.2. Mechanical shaker
4.3. Volumetric flask: 100 ml
4.4. Atomic absorption or equivalent instrumentation

5.0. Reagents

5.1. Sodium acetate 1.0 N buffered to pH 8.2
5.2. Ammonium acetate 1.0 N buffered to pH 7.0
5.3. Isopropyl alcohol: 99%
5.4. Sodium standards in 1.0 N ammonium acetate

6.0. Sample Preparation

6.1. See E&P Sample Preparation

7.0. Procedure

7.1. Weigh 5 g sample into a 50-ml centrifuge tube.

7.2. Add 30 ml of 1.0 N sodium acetate, stopper and shake for 5 min, then centrifuge to clear supernatant.
7.3. Decant and discard supernatant, and repeat step 7.2 three more times to effect sodium saturation.
7.4. Add 30 ml of 99% isopropyl alcohol, stopper and shake for 5 min, then centrifuge to clear supernatant.
7.5. Decant alcohol and discard supernatant, and repeat step 7.4 three more times to effect washing of solids.
7.6. Add 30 ml of ammonium acetate, stopper and shake 5 min, then centrifuge to clear supernatant liquid. Decant supernatant into a 100-ml volumetric flask.
7.7. Repeat step 7.6 two more times decanting into the same volumetric flask.
7.8. Dilute the volumetric to mark with ammonium acetate, and determine sodium concentration by atomic absorption or other instrumentation

8.0. Calculations
8.1. CEC

CEC, meq/100 g = (sodium, meq/litter X 10) / (sample wt., g)

8.2. ESP

$$\text{ESP,\%} = \frac{(\text{Exchangeable Sodium, meq/100g}) \times 100}{(\text{CEC, meq/100g})}$$

9.0. References

9.1. Chapman, H.D. 1965. "Cation Exchange Capacity." C. A. Black (ed.) *Methods of Soil Analysis*. Part 2-Chemical and Microbiological Properties. ASA Agron. Monograph 9. 891-900.

ANNEX F

E&P Sample Preparation

1.0. Scope and Application
1.1. Use this method to prepare samples for analysis by the protocols listed below:
1.1.1. Sodium Adsorption Ratio (SAR)
1.1.2. Exchangeable Sodium Percentage (ESP)
1.1.3. Cation Exchange Capacity (CEC)

2.0. Summary of Method

2.1. Homogenize the sample, dried at 105°C and ground prior to the individual analyses.

3.0. Apparatus and Materials

3.1. Oven capable to 105°C (± 2°C)

3.2. Grinding apparatus

3.3. Drying pans

3.4. Balance

4.0. Procedure

4.1. Homogenize the sample thoroughly.

4.2. Weigh to the nearest 0.1 g a pan large enough to hold 250 g sample.

4.3. Weigh 100 to 200 g homogenized sample to pan, and place pan in oven at 105°C until a constant weight is achieved. Record weights to calculate moisture content.

4.4. Grind the material so it will pass a 2-mm sieve. Sample is now ready for analyses.

5.0. Procedure for Hydrophobic Material

5.1. Tests for hydrophobicity

5.1.1. Visible blobs of oil or grease

5.1.2. Sample presses into a single damp looking mass when crushed with mortar and pestle and will not hydrate with water.

5.1.3. Sample leaves an oily mark when pressed between two pieces of filter paper.

5.1.4. Sample feels damp when pinched between fingers.

5.2. Place sample in muffle furnace and heat to 250°C for 1 hr.

5.3. Raise temperature to 350°C at 50°C intervals allowing smoke to dissipate between adjustments. Do not allow sample to catch fire or exceed 390°C.

5.4. Cool the sample and grind it to pass 2-mm sieve. The sample is now ready for the appropriate analyses.

6.0. Calculation

6.1. Moisture Content (dry weight basis)

Based on weights from steps 4 or 5

$$\text{Moisture,\%} = (W - D)/(D - P) \times 100$$

where W = wet weight of sample + pan, g

D = dry weight of sample + pan, g

P = weight of pan, g

ANNEX G

Exchangeable Cations

1.0. Scope and Application

1.1. This method applies to most soils and E&P waste solids and determines the distribution of cations adsorbed on the solid phase.

2.0. Summary of Method

2.1. The sample is saturated with an excess of ammonium acetate resulting in an exchange of adsorbed cations. The cations released into solution are then quantified as extractable cations and when adjusted for soluble cations are reported as exchangeable cations.

3.0. Interferences

3.1. Sparingly soluble salts may give erroneously high cation distribution values.

4.0. Apparatus and Materials

4.1. Centrifuge and centrifuge tubes

4.2. Mechanical shaker

4.3. Atomic absorption or other suitable instrumentation

5.0. Procedure

5.1. Weigh 5 g of sample to a 50 ml centrifuge tube.

5.2. Add 30 ml 1N ammonium acetate reagent to the tube, stopper, and shake for 5 min and centrifuge to yield a clear supernatant liquid.

5.3. Decant the supernatant as completely as possible into a 100 ml volumetric flask. Repeat step 5.2 two more times combining extracts.

5.4. Dilute to volume (100 ml), mix, and determine the amounts of the various extracted cations using AAS or other suitable instrumentation.

5.5. Soluble cations must be determined for an aqueous extract of the same sample if not determined previously.

6.0. Calculations

6.1. Extractable Cations

extractable cation, meq/100g =

(cation concentration of extract in meq/litre) × 10
(sample wt in g.)

6.2. Soluble Cations
soluble cation, meq/100g = (cation concentration of saturation extract in meq/liter) x (saturation percentage) / 1000

6.3. Exchangeable Cations
exchangeable cation, meq/100g = (extractable cation in meq/100g) (soluble cation in meq/100g)

7.0. References

7.1. Thomas, G.W. 1982. "Exchangeable Cations." A.L. Page (ed.) *Methods of Soil Analysis*. Part 2 - Chemical and Microbiological Properties. 2nd. Edition. ASA Agron. Monograph 9. 159-161.

ANNEX H

Oil & Grease

1.0. Scope and Applications

1.1. Use this method to recover O&G by chemically drying wet E&P waste solids and then extracting by Soxhlet apparatus.

2.0. Summary of Method

2.1. Anhydrous sodium sulfate is used to combine with water and enhance recovery of petroleum hydrocarbon. After drying, the O&G is extracted with methylene chloride using the Soxhlet apparatus.

3.0. Apparatus and Materials

3.1. Soxhlet extraction apparatus
3.2. Analytical balance
3.3. Extraction thimble
3.4. Grease free glass wool
3.5. Vacuum distilling apparatus
3.6. Desiccator

4.0. Reagents

4.1. Concentrated hydrochloric acid
4.2. Anhydrous sodium sulfate
4.3. Nanograde methylene chloride

5.0. Procedure

5.1. Weigh 25 g (±0.5g) of wet E&P waste solid or soil into 150 ml beaker.

5.2. Acidify to pH 2 with concentrated hydrochloric acid.

5.3. Add anhydrous sodium sulfate as necessary to dry solids.

5.4. Transfer sample to extraction thimble, covering sample with glass wool, then place in Soxhlet apparatus.

5.5. Add methylene chloride and commence extraction at 20 cycles/hr for a minimum of 6 hr.

5.6. Using grease free glass wool filter extract into a pre-weighed boiling flask, previously rinsed with solvent.

5.7. Connect boiling flask to vacuum distillation head and evaporate solvent.

5.8. Place boiling flask in a dessicator to cool and remove traces water on glass.

5.9. Weigh boiling flask and record weight gain

6.0. Calculations

6.1. O&G

O&G,% = (weight gain in flask, g) / (sample wt, g) x 100

where:

sample wt. g = (wet weight x 100) / (100 +% moisture)

7.0. References

7.1. Test Methods for Evaluating Solid Waste. 1986. Method 3540. Soxhlet Extraction. EPA SW-846. USEPA. Washington D.C.

APPENDIX II

E&P TOTAL METALS ANALYSIS

INTRODUCTION

The analytical methods presented below follow the Louisiana 29-B protocol, but we modified the procedures to apply to soils and wastes in any locale.

We strongly recommend use of these procedures since the limiting criteria herein contained were derived using the prescribed protocols. Additionally, many of the analytical methods presented here have not been adopted as concise standards. Accordingly, no other documented protocols are available.

ACID DIGESTION OF SLUDGES

Scope and Application

This method is an acid digestion procedure used to prepare sediments, sludges and soil samples for analysis by flame or graphite furnace atomic absorption spectroscopy (FLAA and GFAA, respectively) or by inductively coupled argon plasma spectroscopy (ICP).

Samples prepared by this method may be analyzed by ICP for all the listed metals, or by FLAA or GFAA as indicated below (See Summary of Method). Mercury is analyzed by cold vapor technique.

FLAA	GFAA	HYDRIDE/COLD VAPOR
Barium (Ba)	Arsenic (As)	Arsenic
Cadmium (Cd)	Cadmium	Selenium
Chromium (Cr)	Chromium	Mercury (Hg)
Lead (Pb)	Selenium (Se)	
Silver (Ag)		
Zinc (Zn)		

Summary of Method

A representative 1g sample is digested in nitric acid and hydrogen peroxide. The digestate is then refluxed with nitric acid or hydrochloric acid. Dilute hydrochloric acid is used as the reflux acid for (1) the hydride analysis of As and Se, and (2) the flame AA or ICP analysis of Ba, Ca, Cd, Cr, Pb, and Zn.

Interferences

Sludge samples can contain diverse matrix types, each of which may present its own analytical challenge. Spiked samples and any relevant standard reference material should be processed to aid in determining whether this method is applicable to a given waste.

Hydride Interferences:

1. High concentrations of chromium, cobalt, copper, mercury, molybdenum, nickel, and silver can cause analytical interferences.
2. Traces of nitric acid left following the work-up can result in analytical interferences. Nitric acid must be distilled off by heating the sample until fumes of SO_3 are observed.
3. Elemental arsenic and selenium and many of their compounds are volatile; therefore, certain samples may be subject to losses during sample preparation.

Apparatus and Materials

Conical Phillips beakers: 250-ml
Watch glasses
Thermometer: That covers range of 0 to 200°C
Whatman No. 41 filter paper (or equivalent)
Centrifuge and centrifuge tubes

Reagents

ASTM Type II water (ASTM D-1193): Water should be monitored for impurities.

Concentrated nitric acid, reagent grade (HNO_3): Acid should be analyzed to determine level of impurities. If method blank is <MDL, the acid can be used.

Hydrogen peroxide (30% H_2O_2): Oxidant should be analyzed to determine level of impurities.

Sample Collection, Preservation and Handling

All samples must have been collected using a sampling plan that results in a representative sample.

All sample containers must be prewashed with detergents, acids, and type II water. Plastic and glass containers are both suitable.

Procedure

1.1 Mix the sample thoroughly to achieve homogeneity. For each digestion procedure, weigh to the nearest 0.1 mg and transfer to a conical beaker a 1g portion of sample.

1.2 Add 10-ml of 1:1 HNO_3, mix the slurry, and cover with a watch glass. Heat the sample to 95°C and reflux for 10 to 15 min without boiling. Allow the sample to cool, add 5-ml of concentrated HNO_3, replace the watch glass, and reflux for 30 min. Repeat this last step to ensure complete oxidation. Using a ribbed watch glass, allow the solution to evaporate to 5-ml without boiling, while maintaining a cover of solution over the bottom of the beaker.

1.3 After step 1.2 has been completed and the sample has cooled, add 2-ml type II water and 3-ml of 30% H_2O_2. Cover the beaker with a watch glass and return the covered beaker to the hot plate for warming and to start the peroxide reaction. Care must be taken to ensure the losses do not occur due to excessively vigorous effervescence. Heat until effervescence subsides and cool the beaker.

1.4 Continue to add 30% H_2O_2 in 1-ml aliquots with warming until the effervescence is minimal or until the general sample appearance is unchanged.
Note: Do not add more than 10-ml 30% H_2O_2 total.

1.5 If the sample is being prepared for (a) the hydride analysis of As and Se, or (b) the flame AA or ICP analysis of Ba, Ca, Cd, Cu, Pb and Zn, then add 5-ml of concentrated HCL and 10 ml of Type II water, return the covered beaker to the hot plate, and reflux for an additional 15 min without boiling. After cooling, dilute 100-ml with Type II water. Particulates in the digestate that may clog the nebulizer should be removed by filtration, by centrifugation or by allowing the sample to settle.

a) Filtration: Filter through Whatman No. 41 filter paper (or equivalent) and dilute to 100-ml with Type II water.

b) Centrifugation: Centrifugation at 2,000-3,000 rpm for 10 min is usually sufficient to clear the supernatant of particulates.

c) The diluted sample has an approximate acid concentration of 5.0% (v/v) HCL and 5.0% (v/v) HNO_3. The sample is now ready for analysis.

1.6 If the sample is being prepared for the furnace analysis of As, Cd, Cr, Pb, and Se cover the sample with a ribbed watch glass and continue heating the acid-peroxide digestate until the volume has been reduced to approximately 5-ml. After cooling, dilute to 100-ml with Type II water. Particulates in the digestate should be removed by filtration, centrifugation, or by allowing the particles to settle.

a) Filtration: Filter through Whatman No. 41 filter paper (or equivalent) and dilute to 100-ml with Type II water.

b) Centrifugation: Centrifuge at 2,000-3,000 rpm for 10 min to clear supernatant.

c) The diluted digestate solution contains approximately 5% (v/v) HNO_3. For analysis, withdraw aliquots of appropriate volume and add any required reagent or matrix modifier.

Calculations

The concentrations determined are to be reported on a dry weight basis. A correction is made for the moisture content.

Moisture content is determined on a separate subsample by heating the sample to 105°C.

$$\text{Moisture, \%} = [(\text{wet wt} - \text{dry wt})/\text{dry wt.}] \times 100$$

Quality Control

For each group of samples processed, preparation blanks (Type II water and reagents) should be carried through the entire sample preparation and analytical processes. These blanks will be useful in determining if samples are being contaminated.

Duplicate samples should be processed on a routine basis. Duplicate samples will be used to determine precision. The sample load will dictate the frequency, but 20% is recommended.

Spiked samples and standard reference materials are used to determine accuracy in analysis. A spiked sample should be included with each group of samples processed and whenever a new sample matrix is being analyzed.

The concentration of all calibration standards should be verified against a quality control check sample obtained from an outside source.

References (Acid Digestion of Sludges)

U.S. Environmental Protection Agency. SW-846 2nd Ed., September, 1990. Test Methods for Evaluating Solid Waste. Office of Solid Waste and Emergency Response. USEPA, Washington DC.

TOTAL BARIUM

Scope and Application

This method is used to determine the total amount of barium in soils and solid wastes containing barite.

Summary of Method

The sample is dried and pulverized to pass a 100 mesh sieve and digested in nitric acid. Atomic absorption or ICP emission spectroscopy is used to determine the barium concentration.

Interferences

Since barite (barium sulfate) is only sparingly soluble in acid, the analyst must strictly adhere to the size of the sample and the volume of acid that is specified in the protocol.

Apparatus and Materials

Erlenmeyer flask, 500-ml
Mortar and pestle, agate
Nitric acid (3+7)
100 mesh sieve
Volumetric flask, 500-ml

Procedure

1.1 Prepare the sample as in the sample preparation section.

1.2 Grind a 10g sample to pass a 100 mesh sieve.

1.3 Weigh 100 mg ± 0.1 mg sample into 500-ml flask and add 400-ml Type II water plus 7 ml nitric acid.

1.4 Reflux on a hot plate until the volume in 300-ml

1.5 Transfer to a 500-ml volumetric flask and bring up to the mark.

1.7 Determine the barium concentration by AA or ICP emission spectroscopy.

1.8 Report the barium content as parts per million on a dry weight basis.

APPENDIX III

pH-LIMITING CRITERIA FOR MANAGEMENT OF ONSHORE E&P WASTES

INTRODUCTION

This appendix provides technical justification and guidance for pH limiting criteria for land treating, burial, and roadspreading of Exploration and Production (E&P) wastes.

Resource Conservation and Recovery Act (RCRA) exempts drilling fluids, produced water and associated wastes from being classed as hazardous wastes (40 CFR Part 261). Nevertheless, prudent operation requires knowledge of waste characteristics that make wastes hazardous. Waste pH equal to or less than 2.0 or equal to or greater than 12.5 exhibit the characteristic of corrosivity (40 CFR §261.22(a)(1)). Low pH wastes can react to release from waste toxic levels of heavy metals, which otherwise are insoluble. Also, low and high pH waste pose a threat of chemical burns to humans and animals. E&P operations rely on both caustics and acids. However, less than 0.2 percent of E&P wastes would be classed as corrosive.

The Clean Water Act permit program requires effluent discharges to comply with pH limits, typically within the range 6-9 [40 CFR §133.102(c)]. However, only a limited number of onshore E&P discharges exist. EPA Region VI prohibits all E&P discharges to surface water except Beneficial Use.

This appendix recommends a pH range of 6-9 for road spreading and on-site burial of wastes and a pH range of 6-8 for land treatment of wastes.

pH AND RELATED CHEMISTRY

pH is the measure of hydrogen ion concentration (H^+) of a substance. The negative base 10 logarithm of the (H^+) concentration in moles/liter defines the pH:

$$pH = -\log (H^+) \qquad \text{Eq. A-III-1}$$

This relationship simplifies working with small numbers associated with (H^+) in the natural environment. The range of pH goes from 0 to 14 as shown below:

H^+ moles/liter	10^{-1}	10^{-7}	10^{-14}
pH	1	7	14
reaction	acid	neutral	caustic

Soil reaction expresses the degree of acidity or alkalinity of the soil and is expressed in pH units.

Hydrogen (H^+) and Hydroxide (OH^-) Ion Concentration

Water dissociates into hydrogen and hydroxide ions only slightly. Accordingly, water is known as a weak electrolyte. The dissociation is represented by the following equation:

$$H_2O = H^+ + OH^- \qquad \text{Eq. A-III-2}$$

Pure water at room temperature (32° C) produces an ionization constant at equilibrium (Keq) of 10^{-14}:

$$Keq = \frac{(H^+)(OH^-)}{H^2O} \qquad \text{Eq. A-III-3}$$

Since the activity of water (H_2O) is defined as unity, the equation can be simplified to express the fixed relationship between (H^+) and (OH^-) in pure water:

$$Keq = (H^+)\,(OH^{-1}) = 10\text{-}14 \qquad \text{Eq. A-111-4}$$

with (H^+) and (OH^{-1}) each being 10^{-7} mole/liter. Keq is fixed in that the sum of ion pair is always 10^{-14}. An increase in one ion necessitates a decrease in the other ion. This results in pH shift. For example, sodium hydroxide is a strong base which almost completely dissociates into sodium (Na^+) ions and (OH^{-1}) ions in solution. At 0.1 moles of NaOH /liter, (OH^{-1}) $=10^{-1}$. Since (H^+) (OH^{-1}) = 10^{-14} (Eq. A-III-4), (H^+) equals 10^{-13}. The sodium has no significant effect on either (H^+) or (OH^-) in solution. The pH of the 0.1 moles/liter NaOH solution is:

$$\begin{aligned} pH &= -\log(10^{-13}) \qquad \text{Eq. A-III-5} \\ &= -(-13) \\ &= 13 \end{aligned}$$

Hydrolysis Reactors and Buffers

We define hydrolysis as the reaction of an ion with water to form (H^+) or (OH-). Hydrolysis increases either (H^+) or (OH^-) concentration above the normal 10^{-7} observed for pure water. Cations causing (OH^-) to decrease and (H^+) to increase is defined as acid hydrolysis. Anions causing (OH^-) to increase and (H^+) to decrease is said to undergo base hydrolysis: Aluminum (Al^{+3}) takes part in acid hydrolysis:

$$Al^{+3} + H_2O \longrightarrow Al(OH)^{+2} + H^+ \qquad \text{Eq. A-III-6}$$

since (H^+) increases.

For calcium (Ca^{+2}) the reaction increases (OH^-):

$$CaCO_3 + H_2O \longrightarrow Ca^{+2} + HCO_{3-} + OH^- \qquad \text{Eq. A-III-7}$$

and results in a base hydrolysis.

Anions of strong acids, e.g., (Cl^-) in hydrochloric acid, and cations of strong bases, e.g., Na^+ in sodium hydroxide do not hydrolyze in aqueous solutions. They form neutral salts (pH = 7) when combined, e.g., NaCl.

Ion Exchange

Soil consists of positively and negatively charged particles. Cations in soil solution (hydrogen, H+ aluminum, Al^{+3} calcium, Ca^{+2}) migrate and adsorb to negatively charged soil particles. Soils have a finite capacity to adsorb cations. This capacity is a function of soil type and is reversible, i.e., one cation can be exchanged for another. Thus, we refer this adsorption exchange ability of soils as Cation Exchange Capacity (CEC). The amount and source of negatively charged particles depends on amount of organic matter present, clay content and soil reaction.

Since cations hydrolyze, the amount and speciation of cations (distribution) effects the ability of the soil to consume acid or bases, i.e., buffer capacity.

Soil Buffers

Materials that consume acids or bases while maintaining essentially constant pH are buffers. In soil, buffering can be provided by chemical equilibrium between (H^+) in soil solution (active acidity) or (H^+) or hydroxide cations (Al^{+3} and Fe^{+3}) adsorbed to soil particles (exchangeable acidity) and polymers of iron and aluminum hydroxyl compounds (non-exchangeable acidity). A relatively constant pH results from the release or uptake of these components while maintaining the cation exchange equilibrium.

Cation exchange and hydrolysis provide the basis for buffering in soils. Changes in pH effect microorganisms and some plants. These pH changes result directly from changes in (H^+) concentration or indirectly from chemicals, or nutrient deficiency or imbalances. Buffering capacity of soil pH effectively mitigates these problems.

ANALYTICAL METHODS

pH is measurable for soils and waters. However, except for saturated soils, direct measurement of soil pH at equilibrium is not possible.

Water Analysis for pH

Standard Method (4500-H^{+B}.) commonly used for measuring pH of aqueous solutions relies on a direct reading instrument equipped with a combination glass indicating and calomel reference electrode. EPA guidelines (40 CFR §136, 49 Fed. Reg. 43260) require immediate pH analysis. While immediate analysis is desirable, laboratory determination is more accurate. Assuming the laboratory analysis is performed the same day as the sample is taken, laboratory determinations of pH are preferable.

The direct measurement of solid phase (solids) pH at equilibrium is not possible except under saturated conditions. Saturated soils under field conditions seldom occur, except in wetlands. Accordingly, moisture must be added for analysis. Adding water to the soil produces pH results not representative of interstitial pore liquid in soils at equilibrium.

Adding water to soil results in a metastable condition which changes with time. Accordingly, soil scientists devised a soil pH measurement at thermodynamic equilibrium. The factors affecting thermodynamic equilibria include:

- Spatial or materials variability
- Moisture content or soil/water ratio at which the pH readings are made

Typically, the greater the soil/water ratio, the higher the pH reading (Jackson, 1967). However, drying and grinding soil samples affect reaction kinetics but not the equilibria. Therefore, standardized sample preparation results in reproducible pH measurements, Annex A.

Waste Analysis for pH

Mineral fraction, chemical additives and few soluble salts buffer E&P wastes, i.e., drilling mud. This buffering provides pH stability to the waste. Nevertheless, same day pH analysis assures accurate measurement.

We recommend collecting multiple samples (10 samples) from different areas of the waste area (pit). Composite the samples into one *thoroughly mixed* single sample. This procedure controls sample variability. To assure consistency of measurements, dry (105°C) and lightly grind to pass a 2 mm mesh sieve. Grinding the soil to a fine consistency increases mineral dissolution and shifts equilibria. This results in inaccurate and inconsistent pH measurements.

Moisten a representative sample of ground soil with distilled water to form a saturated paste. Refer to Appendix I, Annex A for Saturated Paste Extract procedure. Measure pH on the saturated paste extract, Appendix III, Annex A.

SOIL BUFFERS

Three conditions describe soil reaction:

- Acidic - pH <7
- Neutral - pH = 7
- Alkaline - pH >7

Precipitation percolation strips basic cations from soil. This process removes basic cations that buffer highly reactive H^+. Alkaline reaction occur in areas of low rainfall and high calcium, magnesium, sodium and potassium content (basic soils). Few soils are neutral (pH = 7), the nearly ideal condition for plant-soil interaction.

Acidic Buffers

Mitra, et al. (1963), Schwertmann and Jackson (1963), Coleman and Thomas (1964), Sawhney and Frink (1966) and Coleman and Thomas (1967) suggest existence of 3 buffer ranges in acid soil:

- Exchangeable H^+ attribute
- Exchangeable Al^{+3} attribute
- Hydrolysis Al^{+3} attribute

Exchangeable H^+ corresponds a pH <4.5 for soil suspensions in water. Exchangeable Al^{+3} occurs between pH 4.0 and 5.5. Hydrolysis of Al^{+3} occurs at pH >5.5 and produces Al $(OH)^{+2}$, $Al(OH)^{+2}$ and polymeric Al-species.

The first and second buffer ranges are dominated by exchangeable Al^{+3} because of extreme reactivity of H^+ and dissolution of soil clay (Thomas, 1960). Non-exchangeable polymeric Al-species form relatively weak buffers and amphoteric in nature (both acidic and basic) and serve as a sink for H^+ and OH^-.

Alkaline Buffers

Alkaline soils predominate in sub-humid and arid regions where insufficient rainfall develops to leach the soil profile (Mehlich, 1943). Soils have pH values from about 6.5 to 7.4 are typically 100 percent base saturated and are considered neutral. These soils do not react to contact with strong acids. Weakly-to-moderately alkaline soils ($7 \leq pH \leq 8.3$) contain free calcium carbonate and react to contact with strong acids.

$$CaCO_3 + HCl \longrightarrow CaCl_2 + H_2O + CO_2 \qquad \text{Eq. A-III-8}$$

Strongly alkaline reaction, i.e., soils having pH > 8.5, result either from a critical percentage of exchangeable sodium (ESP > 15 percent) or free Na_2CO_3 present (US Salinity Lab. Staff, 1954). Dispersion and dissolution of soil humus often accompany soil reactions > pH 9.0 causing the surface to darken, giving rise to the term "black alkali soil" (Fireman and Wadleigh, 1951). Soils remain in a relatively narrow pH range compared with many natural and anthroprogenic products (Fig. A-III-1). Also, soil pH conducive to optimum plant growth generally ranges from 5 to 8.5, with considerable variability noted in both endpoint and width of range between cultivars (Figs. A-III-2, 3 and 4).

pH IMPACTS ON PLANTS

Acidity Impacts

Extensive research has been conducted on the effects of low pH on plant-soil interaction (Hewitt, 1952; Olsen, 1958; Sutton and Halsworth, 1958; Rorison, 1965; Clarkson, 1966; Jackson, 1967).

Direct impacts include:

- Injury by H+
- Physiologically impaired absorption of plant nutrients
- Increased solubility of toxic ions

Soil pH Conditions and Plant Community	**pH Scale**	**Comparative Products**
	0-	
	1-	Hydrochloric acid
		Phosphoric acid
	2-	Lemons
		Vinegar
	3-	Grapefruit
Barren		Apples
		Superphosphate
Extremely Acid Soils	4-	Tomatoes
Suitable for blueberries,		Acid rain
azaleas and rhododendrons		
	5-	Boric acid
Moderately Acid Soils		Fresh beans
Suitable for potatoes,		Distilled water
pine forests and grasses	6-	
Most row crops, vegetables and		Fresh corn
forage crops		Cows milk
Neutral soils	7-	
Alfalfa		Human blood
Slightly alkaline soils		
Hazard of deficiencies of Fe, Mn,	8-	Sea water
Zn and Cu		Calcium carbonate
Strongly alkaline soils		
Adapted species	9-	Bicarbonate of soda
Very strongly alkali soils		
Barren		
	10-	Milk of magnesia
	11-	Ammonia
		Washing soda
	12-	Trisodium phosphate
	13-	Sodium hydroxide
	14-	

Fig. A-III-1 Soil pH conditions and plant communities relative to the pH scale and in comparison to familiar products (Adapted from: W. A. Way, 1968)

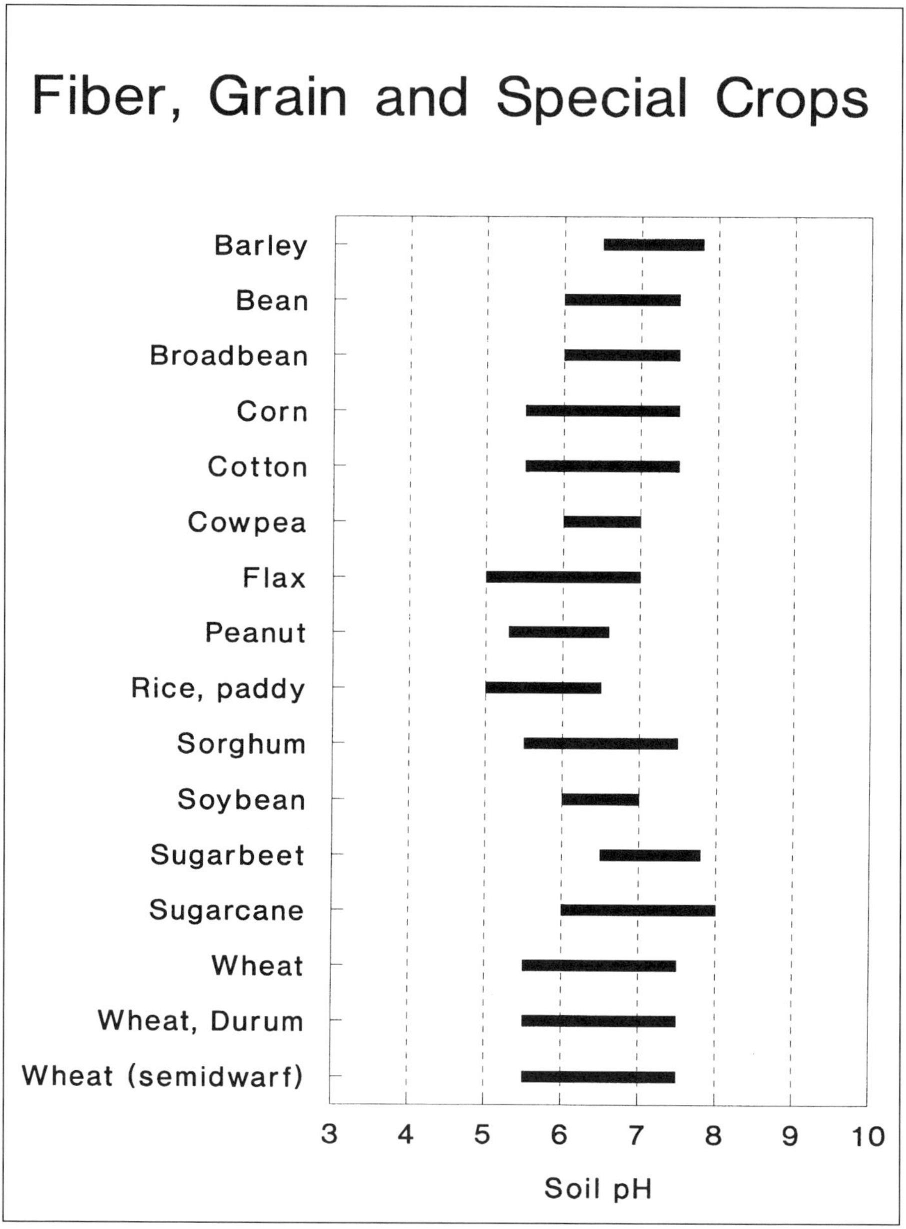

Fig. A-III-2 Soil pH range for optimum plant growth

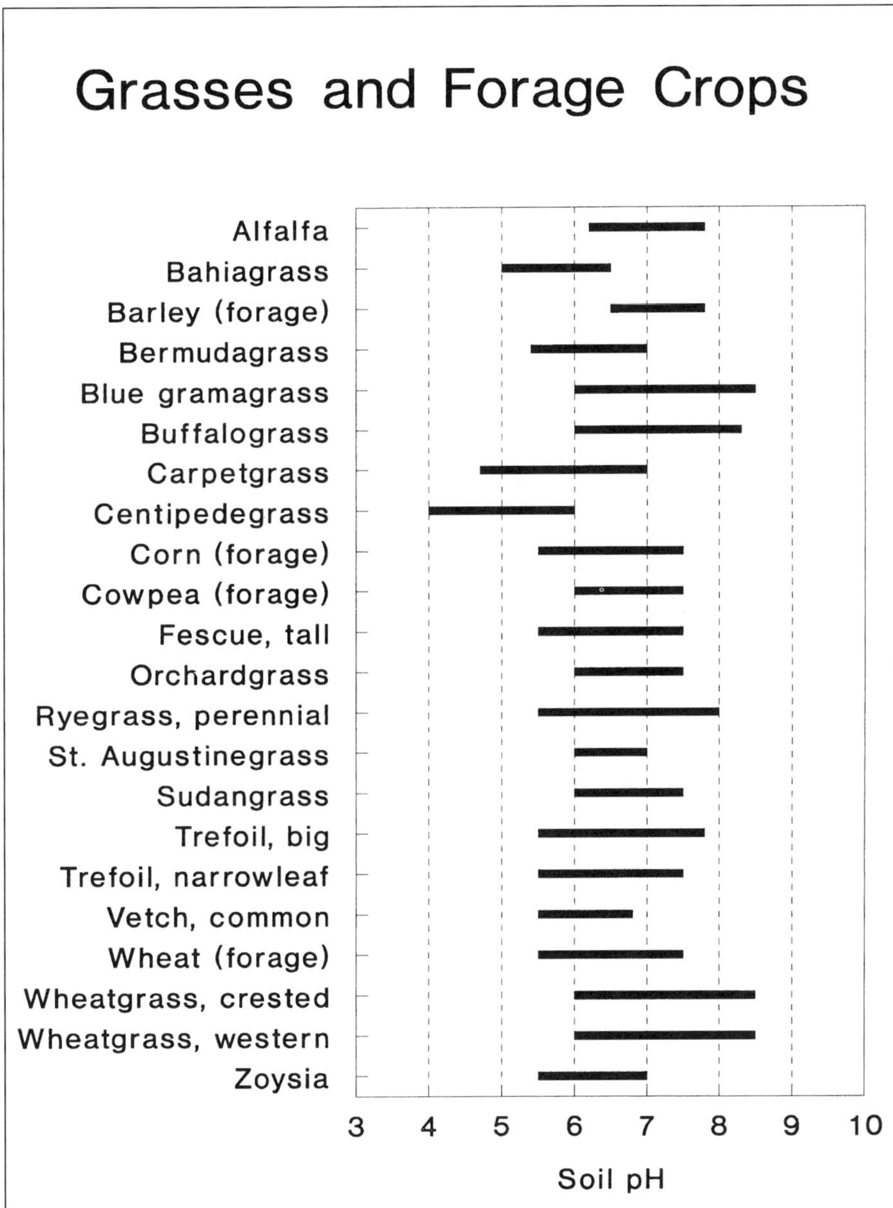

Fig. A-III-3 Soil pH range for optimum plant growth

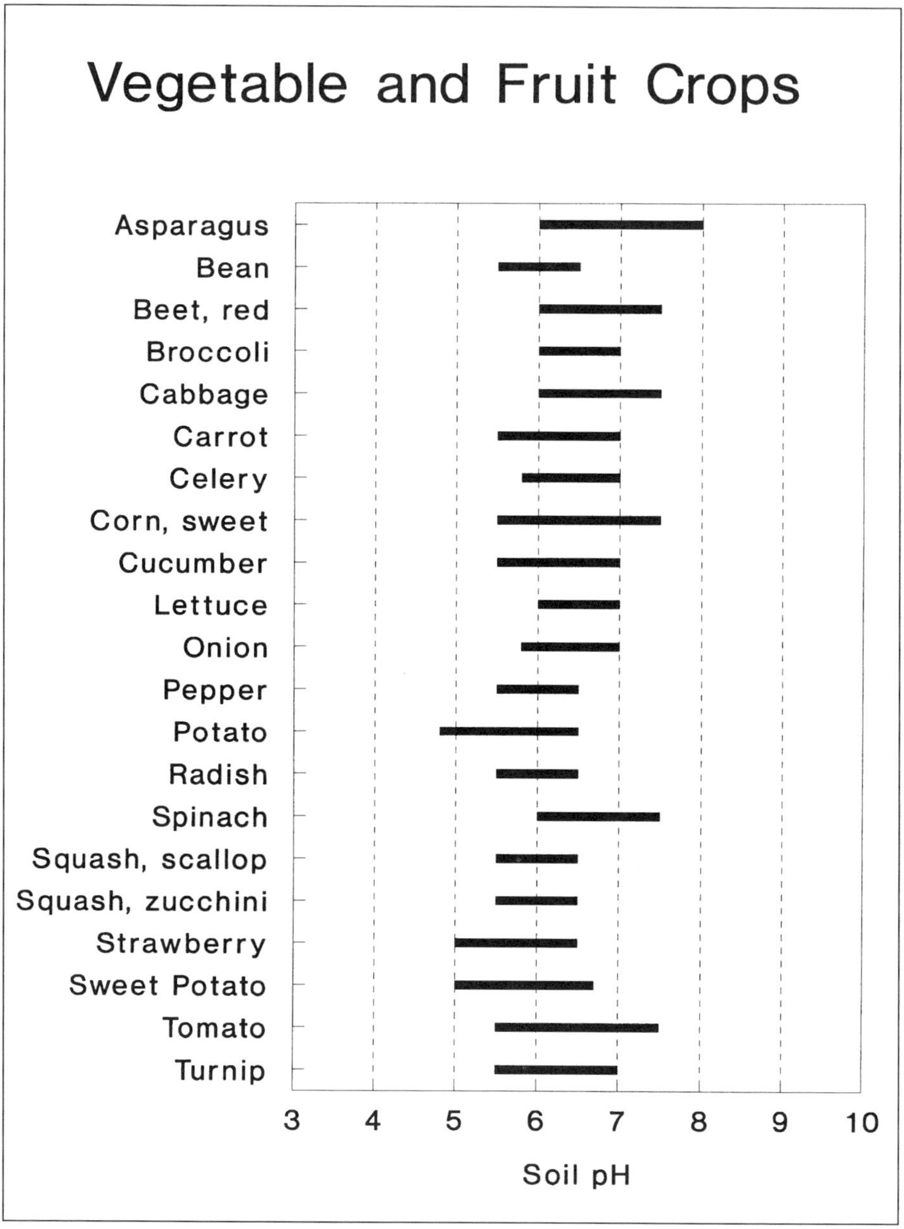

Fig. A-III-4 Soil pH range for optimum plant growth

Injury by H^+, i.e., first buffer range, is short term if the H^+ source is removed or corrected. In the long term and of greater significance is metal toxicity caused by increased solubility as a result of lowered pH and increased acid strength. Metals of concern include aluminum, manganese and iron.

Sparling (1967) developed solubility curves suggesting ferric iron causes no toxicity problem above pH 3.5. Similarly, he found aluminum and manganese typically cause no toxicity problem above pH 5.0. Soil having low oxygen levels i.e., low redox potential, make iron and manganese available in soil. Accordingly, the soil becomes toxic (Gotah and Patrick, 1972). Ferrous iron (Fe^{+2}) becomes soluble at about pH 6.0 under reducing conditions (redox potential E_h <+100 mV). Toxic levels of iron (about 55 mg/l) result in marshland and waterlogged soils. Calcium and magnesium introduce antagonistic mechanisms that make the toxicity threshold for manganese less defined.

Increased solubility and decreased surface adsorption results from increased acidity (Fig. A-III-5). This elevates metal availability in soil solution.

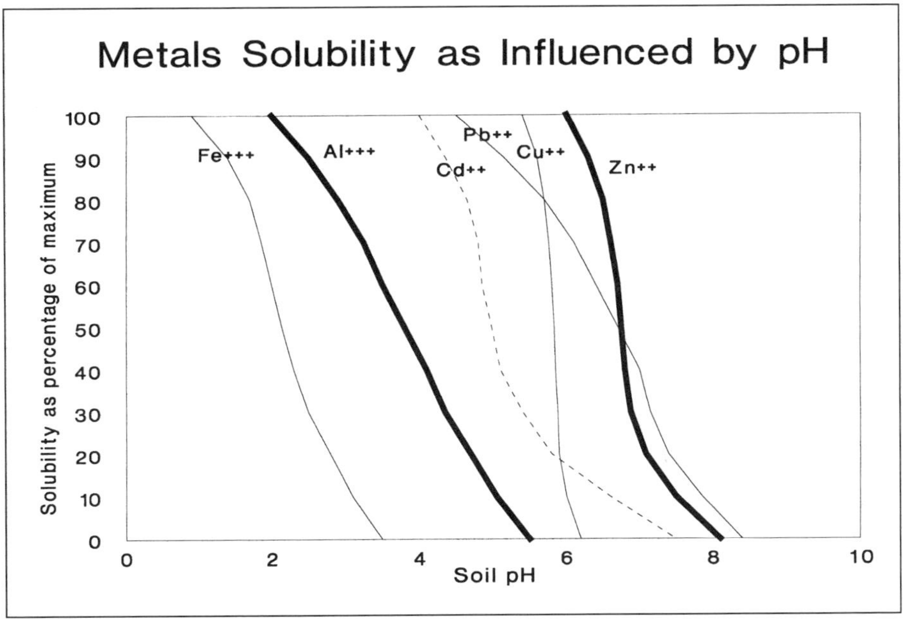

Fig. A-III-5 Metals solubility as Influenced by pH. After Sparling, 1972; Leggett, 1978; Cavarallo and McBride, 1980.

Fertility Impacts

Low pH directly reduces nutrient available in soil. Rorison (1965) and Clarkson (1966) observed the reduced nutrients resulted from precipitation of phosphorous by aluminum in the soil. Similar precipitation occurs within the plant thereby reducing normal phosphorous metabolism.

Indirect impacts of low pH include impaired nitrification (microbiological conversion of ammonium to nitrate) and attach on plants by soil pathogens (Jackson, 1967). Nitrate-nitrogen conversion in soil solution is efficient and rapid compared with ammonium-nitrogen conversion. In fact ammonium-nitrate conversion require roots release hydrogen ions that increases soil acidity.

Alkalinity Impacts

Soil alkalinity can be as devastating to plants as acidity. Additionally, alkalinity is more difficult to correct. High pH reduces available micronutrients (iron, zinc, manganese and copper) to the point plant is undernourished (Hodgson, 1963; Adams, 1965; Grove and Ellis, 1980; Baker and Amacker, 1982). Also, high pH results in loss of nitrogen since under high pH conditions ammonium converts to ammonia gas in the soil (Harsen and Kolenbrander, 1965).

SOIL pH AMENDMENTS

Soil amendments provide means of increasing or decreasing pH.

Decreasing Soil pH

Acidifying amendments neutralize excess alkalinity and lower pH. Aluminum sulfate [$Al_2(SO_4)_3$], ferrous sulfate ($FeSO_4$) and elemental sulfur (S) result in lowered pH. Sulfur works well as a pH reducer. However, because it requires oxidation by soil microorganisms, it works more slowly than the other amendments. Waste acid works well and quickly. However, because of its reactivity, application to effect a desired loading rate into a soil volume of any size in somewhat uniform distribution requires specialized technique and careful management. Additionally, if the acid is truly a "waste" as defined by Resource Conservation and Recovery Act (RCRA), legal difficulties can result from discharging acid to the soil. The amount of acidifying agents is determined empirically by titration to a prescribed endpoint. Soil reaction with acids. Accordingly, rapid determination of the endpoint is essential to avoid overestimating the requirements.

Increasing Soil pH

Lime ($CaCO_3$), the amendment of choice, reduces excess soil acidity. Lime serves two useful functions: (1) neutralize acidity (increases pH) and (2) adds calcium to the soil matrix.

Neutralization of excess acidity occurs at about pH 5.5. Addition of lime in excess of the exchangeable acid becomes necessary since the exchangeable acidity represents that portion of the total acidity replaced by neutral unbuffered salts.

Lime requirements (LR) based on near neutral pH results from adding a contribution for other forms of acidity. Based on resulting pH 6.8, equation A-III-9 provides an estimate of the lime requirement, LR.:

$$\text{LR, meq/100g} = 2(\text{exchange acidity}) + 3[\text{CEC}/(\text{pH})2] \qquad \text{Eq. A-III-9}$$

$$1\ \text{meq/100g} = 1{,}000\ \text{lb. } CaCO_3\text{/acre-6 in.} \qquad \text{Eq. A-III-10}$$

We multiply the exchange acidity by Equation A-III-9 by 2 to compensate for the poor reaction of $CaCO_3$ in soil.

pH CHARACTERISTICS OF E&P WASTES

E&P wastes consist of spent drilling fluids, produced waters and miscellaneous solids. The Resource Conservation and Recovery Act (RCRA) classifies these materials as non-hazardous waste. EPA and API sampled and analyzed these wastes (EPA, 1987a; EPA, 1987b; API 1987) to determine if the exemption should continue.

pH Test Results

EPA and API pH analyses of produced water samples showed E&P liquids as nearly neutral on average Table A-III-1. Reserve pit liquids were moderately alkaline. On the other hand, reserve pit solids show elevated pH levels, Table A-III-1.

One produced water sample measured strongly acidic (EPA pH 3.0 and API pH 3.4). Measured concentrations of iron (2,200 mg/l) and manganese (120 mg/l) strongly suggest the low pH results from oxidation of reduced metals during anaerobic storage. The maximum pH measured by EPA and API (8.1 and 8.7, respectively) reflects bicarbonate/carbonate buffering, even though the produced water is sodium chloride, a neutral salt providing a neutral pH water.

Reserve pit liquids ranged from 6.5 (6.4) to 12.7 (13.0) for EPA (API) analyses, Table A-III-1. The slightly elevated averages pH reflects the standard practice of treating drilling fluids with sodium hydroxide (NaOH) to retain solids in suspension.

Reserve pit solids ranged from 5.9 (7.9) to 12.8 (12.3) for EPA (API) samples. While the mean was 9.5 (10.2), Table A-III-1. The differences between EPA and API pH values result from differences in sampling methods. EPA samples reflect surface conditions where greater potential for fixation of carbon dioxide (Holliday and Deuel, 1990). This naturally acidifying reaction converts hydroxide to a moderately alkaline carbonate buffer system.

Table A-III-1 pH summary statistics for E&P wastes sampled in parallel by EPA and API

	Produced Waters		**Pit Liquids**		**Pit Solids**	
Test Statistic	EPA	API	EPA	API	EPA	API
Sample Size	17	17	15	15	19	19
Average	6.8	6.8	8.2	8.4	9.5	10.2
Median	7.2	6.9	7.6	7.9	9.0	10.3
Mode	7.7	6.8	7.3	7.6	8.8	9.0
Geometric Mean	6.7	6.7	8.1	8.2	9.4	10.2
Variance	1.4	1.2	3.3	3.6	2.8	1.4
Standard Deviation	1.2	1.1	1.8	1.9	1.7	1.2
Standard Error	0.3	0.3	0.5	0.5	0.4	0.3
Minimum	3.0	3.4	6.5	6.4	6.9	7.9
Maximum	8.1	8.7	12.7	13.0	12.8	12.3

Environmental Impact of pH

Produced Water

Little or no short term effects occur when neutral pH produced water contacts soils having pH >5.5. This point emphasizes the importance of immediate soil remediation of produced water releases. Over time, soil pH increases proportionately with the amount of sodium adsorbed onto soil particles. Neutral pH brines contacting acid soils (pH <5.5) typically lowers soil pH 0.5 to 1.2 units more than the reaction from deionized water. As discussed previously, this reaction results from the exchangeable acidity present in the soil and the quantity of salt added by the release. In the long term, displacement of aluminum by sodium raises the pH markedly. Also, plant available phosphorous diminishes as a result of being precipitated by iron and aluminum in the soil.

Near-neutral pH produced water typically result in no adverse impact on pH in alkaline soils over the short term. Weakly calcareous soils can become sodic (ESP ≥15 percent) over the long term, with an anticipated increase in pH of about 2 pH units, as we shift from a $CaCO_3$ to a Na_2CO3 buffered soil. Such a pH increase adversely impacts plants.

Acid water of low ionic strength have an impact similar to acid rain, i.e., decreases pH and increases leaching of nutrient cations in low buffer capacity soils (Galloway, et al., 1978). On the other hand, neutral salt containing acid waters result in little to no adverse short term impact on acid soils. Injury to soils over time is measured by the quantity of sodium uptake. Interestingly, acid brines can have a beneficial impact on plants in calcareous soils when sodium content and ESP are controlled.

Reserve Pit Liquids

Reserve pit liquids contain more alkalinity than produced water. This means pit liquids have a greater potential for adverse impact on weakly buffered soils than produced waters. Application of highly alkaline pit liquids to alkaline soils further hinders plant growth.

On the other hand, controlled applications of alkaline pit liquids to soils having pH <5.5 can be beneficial since excess acidity is neutralized. Also, this process increases the pH dependent cation exchange capacity and increases availability of plant nutrients.

Highly alkaline liquids (pH >12.5) offer a threat to human health and the environment because of the potential to cause caustic burns

Pit Solids

Reserve pit solids (water and oil base muds) analyses show 81.8 percent of the pH values fall between 5 and 8 units, Table A-III-2. The presence of precipitated carbonates and hydroxides account for the strongly buffered solids. Most of the buffers are calcium and magnesium salts, which are not solubilized in the presence of sodium salts in the short term. Even in the long term, adverse impacts of reserve pit solids are low because of leaching of soluble sodium salts. This makes available calcium and magnesium to enhance soil fertility.

Table A-III-2 Frequency distribution for pH of water and oil base drilling muds

Cumulative Interval, pH	Frequency Distribution	Cumulative Percent
< 4.9	34	4.5
5.0 - 5.9	64	13.1
6.0 - 7.9	323	56.1
8.0 - 8.9	226	86.3
9.0 - 9.9	49	92.8
> 10	54	100.0
Total	750	

When applied to strongly acid soil, alkaline drilling fluids solids have a beneficial impact on plant growth because of increased soil pH (Miller and Pesaran, (1980). However, excess amounts of alkaline solids can raise the pH above the acceptable range for plants, i.e., pH >9.

Soil pH values resulting from application of pit liquids and/or pit solids must match the land use (Tables 7-6 and A-I-1). Excess acidity or alkalinity resulting from land disposal of E&P wastes can be corrected by adding lime or elemental sulfur, respectively.

pH CRITERIA

pH 6-9 is satisfactory for wetlands areas but is too high for uplands areas (Tables 7-6 and A-I-1).

pH Management

pH criteria depends on the disposal technique used. The techniques include, (Table A-III-3):

- injection
- road spreading
- land treating

The disposal of E&P waste liquids using annular injection is an option, (Louviere and Reddoch, 1993). Also, injection of liquids is acceptable in Class II wells, 40 CFR Part 146. No pH criteria are imposed on these injections. However, pH adjustment becomes necessary for land treating and road spreading, if the pH criteria is exceeded (Tables 7-6 and A-III-3).

Table A-III-3 Summary of E&P waste, disposal technique, and applicable pH criteria

E & P Waste	Disposal Technique	pH Criteria
Liquids	well injection	NA*
	surface discharge	6 - 9
	road spreading	6 - 9
	land treating	6 - 8
Solids	annular injection	NA*
	land treating	6 - 8
	burial or landfill	6 - 9

*NA - not applicable

Common acidifying agents used for reducing pH include aluminum sulfate ($Al_2(SO_4)_3$) and ferrous sulfate ($FeSO_4$). The agent of choice for increasing pH is calcium carbonate ($CaCO_3$).

The application rate of pH modifying amendments should be based on analytical results. Typically, land spread liquids are treated before application. Follow sampling and compositing recommendation included in Chapter 5. Unfortunately, poorly buffered soils can be over-treated, so it is important to take and analyze samples. Pit solids disposal can be accomplished by (Table A-III-3)

- annular injection
- land treating
- burial or landfill

Annular injection of drill solids is an accepted procedure (Louviere and Reddoch, 1993).

Land treating provides opportunity for oxidation as the result of disking the solids into the soil. Accordingly, sulfur (S), which requires oxidation, is the amendment of choice.

Calcium hydroxide [$Ca(OH)_2$] or sodium hydroxide (NaOH) neutralize excess acidity to be buried or landfilled. Treated acidic waste to be land spread with calcium carbonate ($CaCO_3$).

pH ADJUSTMENT PROCEDURES

Liquids

pH adjustment of liquids is conducted in a tank prior to landspreading or disposal. Several examples of determination of the amount of amendments required are shown below.

Example 1 - Reserve Pit Solids

An operator decides to apply 6,000 barrels of pit liquid to lease roads for dust control. The laboratory reported the pH of the liquids as 11 units. Rapid titration of the composite pit liquids yielded pH 7.0 at alkalinity of 15 meq (OH^-)/l. No other parameter such as salt or oil limits the application of the liquids to the roads. Calculate the quantity of aluminum sulfate ($Al_2(SO_4)_3$) required for alkalinity neutralization as follows:

15 meq(OH^-) /l	=	15 meq H^+/l
$Al_2(SO_4)_3$	=	[(27)x2+{32+16x4}x3] = 342 g/mole
Al has a valence	=	3
g/equivalent	=	342 g/mole/3 eq H^+/mole
$Al_2(SO_4)_3$	=	114 g/equivalent
meq H^+/pit	=	(15 meq H^+/l) × 3.8 l/gal × 42 gal/bbl × 6,000 bbl/pit
	=	14,364,000 meq H^+/pit
since 1 eq H^+	=	1,000 meq H^+
eq H^+/pit	=	14,364,000 meq/pit/(1000 meq H^+/eq H^+)
	=	14,364 eq H^+/pit
amount $Al_2(SO_4)_3$/pit	=	(14,364 eq H^+/pit) x (114 g/eq)/454 g/lb
	=	3,606 lb/pit

Example 2 - Produced Water

An acid frac results in 7,000 barrels of fluid produced into frac tank. The operator decides to apply the liquid to lease roads for dust control. The liquid pH is 3.5 units. Rapid titration of a liquid sample results in 19 meq H^+/liter at pH 7.7. Calculate the amount of calcium carbonate $CaCO_3$ required to neutralize the acidity.

19 meq (H^+)/liter	=	19 meq OH^-/liter
$CaCO_3$	=	[40 + {12 + (16 × 3)}] = 100 g/mole
Ca valence	=	2

g/equivalent	=	(100 g/mole)/(2 eq H^+/mole)
$CaCO_3$	=	50 g/eq.
meq OH^-/tank	=	(19 meq/l) × 3.8 l/gal × (42 gal/bbl) × (7000 bbl/tank)
	=	21,226,800 meq H^+/pit
Again 1 eq H^+	=	1000 meq H^+
eq H^+ pit	=	21,226,800/1000
	=	21,227 eq X OH^-/tank
lb. CaCO3/tank	=	(21,227 eq OH^-/tank × (50 g/eq)/(454 g/lb.)
	=	2,338 lb $CaCO_3$/tank

Solids

Buried or landfilled solids are pH adjusted in the pit prior to disposal. On the other hand, land-spreadable solids are pH adjusted after application to the receiving soil. Neutralization reactions control pH adjustments. Neither acidity nor alkalinity limit the amount of solids. Assume pit solids require 4 acres to manage salt and hydrocarbons. A saturated paste extract pH for the solids equals 11.7 and alkalinity equals 6.7 meq OH^-/100 grams on a dry weight basis. The pit contains 11,500 barrels solids having a moisture content of 255 percent (dry weight basis) and a bulk density equaling 1.55 g/cm^3. The elemental sulfur required to reduce the pH of the solids is calculated as follows:

Sulfur (S)	=	(32 g S/mole)/(2 eq H^+/mole)
	=	16 g S/eq H^+
wet solids, g	=	[1.55 g/cc × 1000 cc/l × 3.8 l/gal x 42 gal/bbl × 11,500 bbl/pit]
	=	2.845×10^9 g/pit
dry solids, g	=	$\frac{(2.845 \times 10^9)(100)}{(100 + 255)}$
	=	8.01 × 108 g/pit
6.7 meq OH^-/100 g	=	0.067 meq H^+/g
meq S/pit	=	(0.067 meq S/g) × (8.01 × 108 g/pit)
	=	(5.367×10^7 meq/pit), but 1 eq =1,000 meq
eq S/pit	=	5.367×10^7 /1000 meq
	=	5.367×10^4 eq/pit
lb. S/pit	=	(5.367×10^4 eq/pit) × (16 g S/eq)/(454 g/lb.)
	=	1891 lb/pit
lb. S/acre	=	(1,891 lb/pit)/(4 acres/pit)
	=	473 lb S/acre

References

Adams, F. 1965. "Manganese." In C. A. Black (ed.) *Methods of Soil Analysis, Part 2.* Chemical and Microbiological Properties. *Agron.*. Am. Soc. Agron., Inc. Madison, WI. 9:1011-1018

API, American Petroleum Institute. 1987. "Oil and Gas Industry Exploration and Production Wastes." DOC No. 471-01-09.

Baker, D. E., and M. C. Amacher. 1982. "Nickel, Copper, Zinc, and Cadmium. *In Methods of Soil Analysis, Part 2.* Chemical and Microbiological Properties. 2nd ed. (ed.) A. L. Page et. al. *Agron.* Am. Soc. Agron., Inc. Madison, WI. 9:323

Cavarallaro, N., and M. B. McBride. 1980. "Activities of Copper and Cadmium in Soil Solutions as Affected by pH." *Soil Sci. Soc.* Am. J. 44:729-732.

Clarkson, D. T. 1966. "Aluminum Tolerance in Species Within Genus Agrostis." *J. Ecol.* 54:167-178.

Coleman, N. T., and G. W. Thomas. 1964. "Buffer Curves of Acid Clays as Affected by the Presence of Ferric Iron and Aluminum." *Soil Sci. Am.* Proc. 28:187-190.

Coleman, N. T., and G. W. Thomas. 1967. "The Basic Chemistry of Soil Acidity." In R. W. Pearsen and F. Adams (eds.): *Soil Acidity and Liming.* Agron. Amer. Soc. Agron., Inc. Madison, WI 12:1-41.

EPA(Environmental Protection Agency). 1987a. "Report to Congress - Management of Wastes from Exploration, Development, and Production of Crude Oil, Natural Gas and Geothermal Energy."

EPA(Environmental Protection Agency). 1987b. Technical Report: "Exploration, Development and Production of Crude Oil and Natural Gas: Field Sampling and Analytical Results." #530- SW-87-005, Appendix A-G.

Fireman, M., and C. H. Wadleigh. 1951. "A Statistical Study of the Relationship Between pH and Exchangeable Sodium Percentage of Western Soils." *Soil Sci.* 71:273-285.

Frink, C. R., and B. L. Sawhney. 1967. "Neutralization of Dilute Aqueous Aluminum Salt Solutions." *Soil Sci.* 103:144-148.

Galloway, J. N., E. B. Cowling, E. Gorham, and W. W. McFee. 1978. "A National Program for Assessing the Problem of Atmospheric Deposition (Acid Rain)." A report to the Council on Environ. Qual. National Atmospheric Deposition Program NC-141. Washington, D.C.

Gotah, S., and W. H. Patrick, Jr. 1972. "Transformation of Manganese in a Waterlogged Soil as Affected by Redox Potential and pH." *Soil Sci. Soc.* Am. Proc. 36:738-742.

Grove, J. H., and B. G. Ellis. 1980. "Extractable Iron and Manganese as Related to Soil pH and Applied Chromium." *Soil Sci. Soc.* Am. J. 44:243-246.

Harmsen, G. W., and G. J. Kolenbrander. 1965. "Soil Inorganic Nitrogen. In W. V. Bartholomew and F. F. Clark (ed.) *Soil Nitrogen. Agron.* 10:43-92.

Helling, C. S., G. Chesters, and R. B. Corey. 1964. "Contributions of Organic Matter and Clay to Soil Cation Exchange Capacity as Affected by the pH of the Saturating Solution." *Soil Sci. Soc.* Am. Proc. 28:517-520.

Hewitt, E. J. 1952. "A Biological Approach to the Problems of Soil Acidity." Trans. Int. Soil Sci. Jt. Meet. Dublin. 1:107- 118.

Hodgson, J. F. 1963. "Chemistry of Micronutrient Elements in Soils." *Advan. Agron.* 15:119-159.

Holliday, G.H., and L.E. Deuel, Jr. (1990) "A Statistical Review of API and EPA Sampling and Analysis of Oil and Gas Field Wastes." Society of Pit Engineers, SPE 20711.

Jackson, W. A. 1967. "Physiological Effects of Soil Acidity." *In Soil Acidity and Liming.* R. W. Pearson and F. Adams (ed.) *Agron.* (Am. Soc. Agron., Inc.) Madison, WI. 12:43-124.

Leggett, G. E. 1978. "Interaction of Monomeric Silicic Acid with Copper and Zinc and Chemical Changes of the Precipitates With Aging." *Soil Sci. Soc.* Am. J. 42:262-267.

Louviere, R.J., and R.J. Reddoch, 1993. "Onsite Disposal of Rig Wastes via Slurrification and Annular Injection."SPE/AIDC 25755.

Mehlich, A. 1943. "The Significance of Percentage Base Saturation and pH in Relation to Soil Differences." *Soil Sci. Soc. Amer.* Proc. 7:167-174.

Miller, R. W., and P. Pesaran. 1980. "Effects of Drilling Fluids on Soil and Plants: II. Complete Drilling Fluid Mixtures." *J. Environ. Qual.* 9:552-556.

Mitra, R. P., B. K. Sharma, and B. S. Kapoor. 1963. "Three Stages in the Titration of Montmorillonite in Water and in Acetonitrile-Benzene Mixtures." *Ind. J. Chem.* 1:225-226.

Olsen, C. 1958. "Iron Uptake in Different Plant Species as a Function of the pH Value of the Nutrient Solution." *Physiol. Pl.* II: 889-905.

Rorison, I. H. 1965. "The Effect of Aluminum on the Uptake and Incorporation of Phosphate by Excised Sainfoin Roots." *New Phytol* 64:23-27.

Sparling, J. H. 1967. "The Occurrence of Schoenus Nigricans L. in Blanket Bogs. II. Experiments on the Growth of S. Nigricans Under Controlled Conditions." J. Ecol. 55:15-31.

Sutton, C. D., and E. G. Hallsworth. 1958. "Studies on the Nutrition of Forage Legumes. I. Toxicity of Low pH and High Manganese Supply to Lucerne, as Affected by Climatic Factors and Calcium Supply." Pl. Soil. 9:305-317.

Schwertmann, U.,and M. L. Jackson. 1963. "Hydrogen-Aluminum Clays: A Third Buffer Range Appearing in Potentiometric Titration." Sci. 139:1052-1053

Soil Salinity Lab. Staff. 1954."Diagnosis and Improvement of Saline and Alkali Soils." USDA Handbook 60. U.S. Printing Office, Washington, D.C.

Standard Methods for the *Examination of Water and Waste Water*, 18th Ed. 1992.Method 4500-H+B. Electrometric Method.

Way, W. A. 1968. "The Whys and Hows of Liming." U of Vt. Brieflet 997.

ANNEX A

pH Determination

1.0. Scope and Application

- 1.1. This method is an electrometric procedure that provides a direct measurement of pH in water, soils and E&P wastes.

2.0. Summary of Method

- 2.1. The pH of E&P liquids and solids are measured in the field whenever possible.
- 2.2. The dried and prepared sample (soil or solid waste) is mixed with type II water.

3.0. Interferences

- 3.1. Samples with very low or very high pH may give incorrect readings on the meter.
- 3.2. Temperature fluctuation will cause measurement errors.
- 3.3. Errors will occur when the electrodes become coated with oily materials. The electrode can either (1) be cleaned with an ultrasonic bath, or (2) be washed with detergent, rinsed with water, followed by 1:10 HCl, then deionized water.

4.0. Apparatus and Materials

- 4.1. Temperature compensating pH meter.
- 4.2. Electrodes:
- 4.2.1. Calomel reference electrode.
- 4.2.2. Glass indicating electrode.
- 4.2.3. Or a combination glass-calomel electrode.

4.3. 50 ml beaker or other suitable container.

5.0. Reagents

5.1. ASTM Type II water (ASTM Dl193).

5.2. Standard buffers, usually pH 2, 4, 7, 10 and 12.

6.0. Sample Preparation

6.1. See sample preparation section (Annex D).

7.0. Procedure

7.1. Calibration:

7.1.1. Set up instrument according to the manufacturers instruction. Operational function and settings are to varied to be covered in this generalized method.

7.1.2. Standardize instrument/electrode systems against buffers that bracket the expected pH of the sample(s).

7.2. pH measurement of samples:

7.2.1. Insert electrodes into aqueous solutions or E&P liquids and read pH immediately on the standardized system.

7.2.2. Mix or stir E&P waste suspensions to be measured in the field or as received for 10 sec, then insert electrodes and read pH immediately on the standardized system.

7.2.3. Stir saturated paste for 10 sec, then insert electrodes and read pH immediately on the standardized system. A glass rod attached to the electrodes or tip cover will serve to protect electrodes, as will a slight swirling action, on insertion.

8.0. Quality Control

8.1. Duplicate samples and check standards should be analyzed routinely.

8.2. Rinse electrodes between samples.

9.0. References

9.1. EPA method 9045. Test Methods for Evaluating Solid Waste-Physical/Chemical Methods. EPA SW-846, 1986, revised.

9.2. Method 12-2.6. Glass Electrode-Calomel Electrode pH Meter Method. Methods of Soil Analysis. Part 2-Chemical and Microbiological Properties. A. L. Page (ed.) Agron. 9. Amer. Soc. Agron. Madison, WI. 1982.

ANNEX B

Alkalinity (Acid Requirement)

1.0. Scope and Application

1.1. Alkalinity is a measure of the base forms in E&P waste liquids or solids that are readily neutralized on titration with a strong acid. The acidity requirement is a measure of the amount of acid needed to lower the waste matrix pH to < 8.0.

2.0. Summary of Method

2.1. Alkalinity is measured by potentiometric titration with hydrochloric acid (HCl) used as the titrant and reported in meq OH^-/liter for liquids or meq OH^-/100 g for solids.

2.2. E&P waste liquids are titrated direct with a pH meter and appropriate electrodes.

2.3. E&P waste solids are titrated in-situ after preparing a dilute suspension with Type II water.

3.0. Interferences

3.1. Exposure to air and fixation of CO_2. This is overcome by rapid titration to pH 6.5.

4.0. Apparatus

4.1. Temperature compensating pH meter

4.2. Electrodes

4.2.1. Calomel reference electrode

4.2.2. Glass electrode

4.2.3. Or a combination glass-calomel electrode

5.0. Reagents

5.1. ASTM Type II water (ASTM D 1193)

5.2. pH 4 and pH 10 calibration buffer solutions

5.3. 0.1 N hydrochloric acid (HCl)

5.4. 50 ml burette

6.0. Procedure

A. E&P Waste Liquids

6.1. Titrate a suitable aliquot of E&P waste liquid such that a minimum 20 ml of the 50 ml burette is utilized to the potentiometric endpoint (pH 6.5).

6.2. The pH is monitored on a direct reading instrument with sample under constant agitation while adding acid.

B. E&P Waste Solids

6.1. Weigh 10 g E&P waste solid sample into 250 ml wide mouth erlenmeyer flask or tall form beaker.

6.2. Add 150 ml Type II water, and insert electrodes for immediate titration.

6.3. Titrate under constant agitation to potentiometric endpoint (pH 6.5).

7.0. Calculations

7.1. Alkalinity (Acid Requirement)

a. E&P Liquid (meq OH-/liter)

meq OH-/liter = A x B x 1000 / ml sample

where :

A = ml titrant used

B = normality (N) of titrant

b. E&P Solid (meq OH-/100 g)

meq OH-/100 g = $A \times B \times 100$ / g sample

where :

A = ml titrant used

B = normality (N) of titrant

g sample (dry) = [g sample (wet) × 100] / [100 + % moisture]

ANNEX C

Active Acidity (Lime Requirement)

1.0.Scope and Application

1.1. Active acidity is a measure of the acid forms in E&P waste liquids or solids that cause elevated hydrogen ion (H^+) activities and pH values < 6.0. The lime or base requirement is a measure of the amount of base needed to raise the pH to > 6.0.

2.0. Summary of Method

2.1. Active acidity is measured by potentiometric titration using sodium hydroxide as the titrant and reported in meq H^+/liter for liquids or meq H^+/100 g for solids.

2.2. E&P waste liquids are titrated direct with a pH meter and appropriate electrodes.

2.3. E&P waste solids are titrated in-situ after preparing a dilute suspension with Type II water.

3.0. Interferences

3.1. Aluminosilicate minerals associated with E&P waste solids can consume erroneously high quantities of base by neutralization of polymerized species and/or formation of aluminates when titrated too slowly. This is overcome by rapid titration and holding the pH 7.5 endpoint 30 sec.

4.0. Apparatus

4.1. Temperature compensating pH meter.

4.2. Electrodes

4.2.1. Calomel reference electrode.

4.2.2. Glass electrode.

4.2.3. Or a combination glass-calomel electrode.

5.0. Reagents

5.1. ASTM Type II water (ASTM D 1193).

5.2. pH 4 and pH 10 calibration buffer solutions.

5.3. 0.1 N sodium hydroxide (NaOH).

5.4. 50 ml burette.

6.0. Procedure

A. E&P Waste Liquids

6.1. Titrate a suitable aliquot of E&P waste liquid such that a minimum 20 ml of the 50 ml burette is utilized to the potentiometric endpoint (pH 7.5).

6.2. The pH is monitored on a direct reading instrument with sample under constant agitation while adding base.

B. E&P Waste Solids

6.1. Weigh 10 g E&P waste solid sample into 250 ml wide mouth erlenmeyer flask or tall form beaker.

6.2. Add 150 ml Type II water, agitate and allow to stand 30 min.

6.3. Insert electrode(s)and titrate under constant agitation to potentiometric endpoint (pH 7.5).

7.0. Calculations

7.1. Active Acidity (Lime Requirement)

a. E&P Liquid (meq H^+/liter)

meq H^+/liter = $A \times B \times 1000$ / ml sample

where :

A = ml titrant used

B = normality (N) of titrant

b. E&P Solid (meq H^+/100 g)

meq H^+/100 g = $A \times B \times 100$ / g sample

where :

A = ml titrant used

B = normality (N) of titrant

g sample (dry) = [g sample (wet) × 100] / [100 + % moisture]

ANNEX D

E&P Sample Preparation

1.0. Scope and Application

1.1. This method is used to prepare samples for analysis by the protocols listed below:

1.1.1. Saturated Paste Extract

1.1.2. pH

1.1.3. Active Acidity (Lime Requirement)

1.1.4. Alkalinity (Acid Requirement)

2.0. Summary of Method

2.1. The sample is homogenized, dried at 105°C and ground prior to the individual analyses.

3.0. Apparatus and Materials

3.1. Oven capable to 105°C (+/- 2°C)

3.2. Grinding apparatus

3.3. Drying pans

3.4. Balance

4.0. Procedure

4.1. Homogenize the sample thoroughly.

4.2. Weigh a pan to the nearest 0.1 g that is large enough to hold 250 g sample.

4.3. Weigh 100 to 200 g homogenized sample to pan, and place pan in oven at 105°C until a constant weight is achieved. Record weights to calculate moisture content.

4.4. Grind the material so that it will pass a 2-mm sieve. Sample is now ready for the analyses listed.

5.0. Procedure for Hydrophobic Material

5.1. Tests for hydrophobicity

5.1.1. Visible blobs of oil or grease

5.1.2. The sample presses into a single damp looking mass when crushed with mortar and pestle and will not hydrate with water.

5.1.3. Sample leaves an oily mark when pressed between two pieces of filter paper.

5.1.4. Sample feels damp when pinched between fingers.

5.2. Place sample in muffle furnace and heat to 250°C for 1 hr.

5.3. Raise temperature to 350°C at 50°C intervals allowing smoke to dissipate between adjustments. Do not allow sample to catch fire or exceed 390°C.

5.4. Cool the sample and grind it to pass 2-mm sieve. The sample is now ready for the appropriate analyses.

6.0. Calculation

6.1. Moisture Content (dry weight basis)

$$\text{Moisture, \%} = (W - D)/(D - P) \times 100$$

where

W = wet weight of sample + pan, g

D = dry weight of sample + pan, g

P = weight of pan, g

APPENDIX IV

PIT CLOSURE DESIGN

from Guidelines for Closing Drilling Waste Fluid Pits in Wetlands and Upland Areas (Freeman and Deuel, 1984)

A total of 293 pits were closed in Louisiana and Mississippi. These pits were sampled and evaluated for compliance with GLC values, and closed utilizing the design procedures formulated in this Appendix.

The design procedures were summarized in prognosis form for personnel to utilize during closure operations. Field personnel prepared an individual report for each pit closure operation. This report was placed on file at the office for closure verification.

The RCRA amendments require operators to record the legal location of pits to be closed and to identify the composition of waste fluids. Records of legal locations, waste fluids analysis and pit closure designs are being maintained to comply with RCRA 1980 amendments.

1.1 CLOSURE DESIGN PROCEDURES

Four basic closure design methods are utilized when closing pits in wetland or upland areas - Minimum Management, Metal Management, Organic Management and Salts Management. In addition, these design methods take into account whether pits are located in submerged or elevated areas.

In submerged wetlands the pit area will be inundated with water after the pit is closed. This type closure design is more difficult to complete, because dilution of pit wastes with surface soils is limited, as the only soil available for mixing is pit levee materials. Consequently, the reduction of waste constituent concentrations to GLC values through surface soil dilution is limited. If waste constituent concentrations exceed GLC values after closure, by more than "factors" of 2-3, waste fluids must be removed for disposal off site. These "factors" do not apply to organics which are limited to 5% by weight and biodegraded prior to pit closure.

In elevated wetlands reduction of constituent concentrations can be achieved in most cases by utilizing surface soil dilution. The waste fluids can be thoroughly mixed with pit levees and adjacent soil at the well location. Waste constituent concentrations can exceed GLC "Maximum Values Before Closure" at elevated wetlands as long as sufficient surface soil is available for dilution to GLC values after closure. Otherwise, the waste fluid must be injected underground or transported off site for disposal.

In upland areas the reduction of constituents can be achieved easily by utilizing surface soil mixing. However, land treatment of waste fluids may be difficult to do at upland areas, because landowners are reluctant to give permission to spread wastes onto adjacent property. Burial of waste fluids in trenches may be the only on site disposal option available. At present, regulations have not been issued for burying waste fluids in Louisiana, Mississippi or Alabama. However, waste fluids should be mixed with soil during burial to reduce metal concentrations to GLC. If metal concentrations cannot be diluted to GLC during burial, low permeability soil should be placed over trenches to prevent leaching of metals to groundwater. If groundwater levels are shallow, operators should consider removing waste fluids for underground injection or off site disposal. Solidification of waste fluids is another disposal alternative, but is beyond the scope of this report.

1.1.1 Minimum Management

Pits containing waste fluid constituent concentrations below GLC values can be closed with minimum management. Minimum management also applies to waste fluids with metal concentrations below "maximum values before closure". Before closure values were formulated for pits located in submerged areas, but can apply to all pits prior to closure for management design decisions.

Pit liquids must be removed prior to closure. Injecting liquids underground or transporting liquids off site increase closure costs. In Mississippi liquids can be discharged off site to waters of the United States. Pit liquids, in most cases, can be treated to comply with effluent limitations. Suspended solids, metals and pH usually can be altered to meet effluent limitations by utilizing lime treatments, Section 1.2, of this Appendix. This substance can be used to flocculate suspended solids and/or adjust pH to desired levels. Adjusting pH also removes metals from liquids through precipitation

Pits with saline liquids located in freshwater wetlands or uplands must be diluted with freshwater to comply with effluent limitations. Otherwise, saline liquids cannot be discharged off site and must be injected underground or removed for off site disposal in accordance with state regulations. Discharging saline pit liquids to brackish wetlands will not normally present a problem; however, the state agency has discretion in approving such discharges.

After pit liquids are removed, pit waste fluids must be thoroughly mixed with pit levee materials. In elevated wetland or upland areas, waste/soil mixture ratios should be high enough to support the weight of vehicles or facility equipment. Closure areas are graded to natural wetland elevations. Closure areas located in elevated wetland or upland areas should be treated with 50 lbs./acre of balanced fertilizer (10-10-10 or 12-12-12) to complete closure.

1.1.2 Metal Management.

As discussed in 1.1.1 above, pit liquids must be analyzed and permission obtained from state agencies before being discharged into waters of the United States. If liquids cannot be treated or discharged under state regulations, liquids must be removed and disposed of in accordance with applicable regulations.

1.1.2.1 Submerged Wetland Areas.

If metal concentrations in waste fluids are greater than GLC maximum values before closure, the required volume of waste fluid must be removed from the pit to assure the final waste/levee mixture will comply with GLC values after closure. (Waste fluids removed must be disposed of in accordance with state or federal regulations.) For example, if an LC value of zinc is 2,000 ppm in the waste fluid, the waste/levee mixture should reduce the zinc concentration to approximately 1,000 ppm. The GLC value set for zinc in the final waste/levee mixture after closure is 500 ppm. Consequently, 75% of pit materials must be removed to achieve the final waste/levee mixture of 500 ppm, taking into consideration that the levee materials contain about 125 ppm zinc. The pH of the residual waste fluid should be adjusted between 6.5 to 7.5 prior to mixing with levee materials, Section 1.2 of this Appendix. Treated waste fluids and levee materials should be mixed thoroughly. The area is then graded to the natural wetland elevation.

1.1.2.2 Elevated Wetland Areas.

Metal management at elevated wetland areas is easier to accomplish than at submerged areas because adjacent soil is available for mixing. Two closure options assure protection of the wetland environment.

1.1.2.2.1

Metal concentrations in waste fluids exceed GLC maximum values before closure, and sufficient land mass is not available to dilute metals to GLC values. Under this option required waste fluids must be removed (see calculation in 1.1.2.1) from the pit and either injected underground or transported off site to a commercial disposal facility (waste fluids removed must be disposed of in accordance with state and federal regulations).

Remaining waste fluids must be removed from the pit and spread over the well location. Waste fluid thickness should not exceed four inches and should be allowed to dry before mixing with surface soil. The pH of the waste/levee mixture can be adjusted by spreading lime or sulfur over the dried waste material and disking the waste and lime into the surface soil to a depth of six inches, Section 1.2 of this Appendix. The adjusted pH value should range from 6.5-7.5. Grade treatment area to natural wetland elevation and treat with 50 lbs./acre of a balanced (10-10-10 or 12-12-12) fertilizer to complete closure.

1.1.2.2.2

Metal concentrations in waste fluids exceed GLC value before closure and sufficient land mass is available to dilute metals to GLC values. Under this option, waste fluids must be removed, spread and treated on site as described in Paragraph 1.1.2.2.1 above.

1.1.2.3 Upland Areas.

Landowners are sensitive to operators spreading waste fluids onto adjacent property. Operators may be required to bury or transport waste fluids off site for disposal. Prior to burial or spreading of waste fluids, operators should inject underground as much of the waste fluids as possible, if permission from state agencies can be obtained. Two closure options assure protection of the environment.

1.1.2.3.1

Metal concentrations in waste fluids exceed GLC maximum values before closure and sufficient land mass is not available to dilute metals to GLC values. Waste fluids should be mixed with soil during burial to reduce metals to GLC. The pH of the waste/soil mixture during burial should range from 6.5-7.5. If metal concentrations cannot be reduced to GLC (but will not exceed TCLP toxicity limitations), a low permeability soil should be placed over burial trenches to prevent leaching of constituents to groundwaters. If groundwater levels are shallow, residual waste fluids should be transported off site for disposal. Grade closure area to natural upland elevation and treat with 50 lbs./acre of balanced (10-10-10 or 12-12-12) fertilizer to complete closure.

1.1.2.3.2

Metal concentrations in waste fluids exceed GLC maximum values before closure and sufficient land mass is available to dilute metals to GLC values. Waste fluids should be mixed with soil during burial to reduce metals to GLC. The pH of the waste/soil mixture during burial should range from 6.5-7.5. If metal concentrations cannot be reduced to GLC with trenching soils, or if groundwater is shallow, waste fluid should be spread over adjacent land area to reduce metals to GLC. Waste fluid thickness should not exceed four inches and should be allowed to dry before mixing with surface soil. The pH of the waste/soil mixture should be adjusted by spreading lime or sulfur over dried waste material and disked soil to a depth of six inches, Section 1.2 of this Appendix. Adjusted pH value should range from 6.5-7.5.

After burying or disking is complete, grade closure area to natural upland elevation and treat with 50 lbs./acre of a balanced (10-10-10 or 12-12-12) fertilizer to complete closure.

1.1.3 Organic Management.

High concentrations of organics adversely impact biological and plant life. GLC values are shown in Table A-I-1. Closure of pits to meet GLC values is sometimes difficult due to recalcitrance of organics. If oil and grease levels are in excess of 5% for submerged wetlands or 10% for elevated wetlands or uplands, high oil and grease levels can require 1-2 years of mechanical aeration and nutrient balancing for decomposition to GLC.

As discussed in 1.1.1 above, pit liquids must be removed prior to closure. Liquids must be analyzed and permission obtained from appropriate agencies before being discharged onto adjacent property. If liquids cannot be discharged onto adjacent property, liquids must be removed and disposed of in accordance with state or federal regulations.

1.1.3.1 Submerged Wetland Areas.

Biodegrading organics in submerged wetland areas is difficult and time consuming. Waste fluids containing in excess of 5% organics may require several years of mechanical aeration and nutrient balancing for decomposition to GLC values. Consequently, waste fluids should be either injected underground or transported off site to a commercial facility. Closure options to meet GLC values and assure protection of submerged wetland environments are dependent on oil and grease concentration.

1.1.3.1.1

Oil and grease concentrations in waste fluids are between 1 and 5%. Mix waste fluids thoroughly with nitrogen fertilizers to biodegrade organic levels to GLC value before closure. Nitrogen must be available to biota for degradation to occur. Mix sufficient ammonium sulfate [$(NH_4)_2SO_4$] with waste fluid to provide a carbon to nitrogen ratio of 150:1, Section 1.3 of this Appendix. Successive applications of [$(NH_4)_2SO_4$] may be required to assure sufficient biodegradation. Complicated treatments should be initiated beginning sometime in late spring because increased temperature enhances degradation rates. Pit waste materials should not be mixed with pit levee materials until degradation has reduced oil and grease to GLC. Once analyses showing degradation to GLC has occurred, mix pit waste materials thoroughly with pit levee materials and grade pit area to natural wetland elevation.

1.1.3.1.2

Oil and grease concentrations in waste fluids are greater than 5%. Remove waste fluids from pit by injecting underground or transporting off site. (Waste fluids must be disposed of in accordance with state or federal regulations.) After waste fluids are removed, mix residual waste fluids thoroughly with pit levee materials. Residual waste fluid/levee material mixture should not exceed 1% oil and grease after closure. Grade closure area to natural wetland elevation.

1.1.3.2 Elevated Wetland Areas.

Organic management at elevated wetland areas provide a much better opportunity for biodegrading waste fluids to GLC. However, waste fluids containing oil and grease concentrations in excess of 10%, may require several years of mechanical aeration and nutrient balancing for decomposition to GLC. Consequently, waste fluids containing organics >10% must be either injected underground or transported off site to a commercial facility. Three closure options assure protection of the elevated wetland environment.

1.1.3.2.1

Oil and grease concentrations in waste fluids are between 1-10% and sufficient land mass is available to spread waste fluids. Spread waste fluids over land surface not to exceed four inches in thickness. Grade treatment area to natural wetland elevation.

Once waste fluids have dried, adjust carbon to nitrogen ratio (150 to 1) with [$(NH_4)_2SO_4$], and disk to a depth of six inches (see Section 1.4.3, of this Appendix). If waste/soil mixture exceeds 3% oil and grease, spread additional [$(NH_4)_2SO_4$] applications at eight and 16 week intervals and disk treated area to six inches. Include a balanced (10-10-10 or 12-12-12) fertilizer treatment of 50 lbs./acre with last ammonium sulfate treatment to complete closure.

1.1.3.2.2

Oil and grease concentrations in waste fluids are between 1-10% and land mass is not available to spread waste fluids. Remove required volume of waste fluids from pit by injecting underground or transporting offsite (waste fluids must be disposed of in accordance with state or federal regulations).

Required volume for removal can be determined by (1) calculating volume of waste fluids to be spread over available area (maximum spreading thickness is four inches) and (2) subtracting calculated volume from total pit volume. Once required volume of waste fluid is removed from pit, spread remaining waste fluid onto adjacent land area not to exceed a thickness of four inches. Once waste fluids dry, treat and disk area as described in 1.1.3.2.1 above.

1.1.3.2.3

Oil and grease concentrations in waste fluids are greater than 10% and sufficient land mass is not available to spread wastes. Land necessary for biodegrading oil and grease concentrations to GLC can be determined easily. First, calculate waste/soil dilution ratio necessary to reduce waste fluid oil and grease concentrations $< 10\%$. For example, an operator wants to land treat a waste fluid containing 40% oil and grease. One inch of waste fluid mixed with five inches of native soil will reduce oil and grease concentration to 6.7% (10% is necessary to biodegrade organics in one season). Secondly, calculate land area necessary for diluting 40% oil and grease waste fluids to 6.7%. Knowing that only one inch of 40% waste can be spread onto adjacent land and knowing the volume of waste fluids in the pit, the required volume for removal can be determined.

Remove excess waste fluids from pit by either underground injection or transportation off site (waste fluids must be disposed of in accordance with state or federal regulations). Once required volume of waste fluid is removed from pit, spread remaining waste fluid onto adjacent land area at required waste/soil dilution ratio. Once waste fluids dry, treat and disk area as described in 1.1.3.2.1 above.

1.1.3.2.4

Oil and grease concentrations in waste fluids are greater than 10% and sufficient land mass is available to spread wastes. Calculate land area necessary for biodegrading waste fluids as described in Section 1.1.3.2.3 above.

Spread waste fluids onto calculated land area at required waste/soil dilution ratio. Once waste fluids dry, treat and disk area as described in 1.1.3.2.1 above.

1.1.3.3 Upland Areas.

Organic management at upland areas will be difficult to initiate because landowners may not give permission to spread wastes onto adjacent properties. Operators may be required to bury waste fluids on site or transport waste fluids off site for disposal. Prior to burial or spreading of waste fluids, operators should inject as much of waste fluids as possible underground, unless permission from state agencies cannot be obtained. Three closure options assure protection of the land environment.

1.1.3.3.1

Oil and grease concentrations in waste fluids are greater than 1% and sufficient land mass is not available to spread waste fluids. Waste fluids should be mixed with soil during burial to reduce oil and grease concentrations to GLC. If oil and grease concentrations cannot be reduced to GLC, low permeability soil should be placed over burial trenches to prevent leaching of constituents to groundwater. If oil and grease concentration of waste/soil mix cannot be reduced below 10% prior to burial or groundwater is shallow, operator should consider transporting wastes off site to a commercial facility. Grade closure area to the natural upland elevation and treat with 50 lbs./acre of balanced (10-10-10 or 12-12-12) fertilizer to complete closure.

1.1.3.3.2

Oil and grease concentrations in waste fluids are between l-10% and sufficient land mass is available for spreading. Waste fluids should be spread over adjacent land and treated as described in Section 1.1.3.2.1.

1.1.3.3.3

Oil and grease concentrations in waste fluids are greater than 10% and sufficient land mass is available for spreading. However the option to land spread is generally exercised after one rules out burial. Waste fluids should be mixed with soil during burial to reduce oil and grease concentrations below 10%. Low permeability soil should be placed over burial trenches to prevent leaching of constituents to groundwater.

If oil and grease concentrations of waste/soil mix cannot be reduced below 1O% prior to burial or groundwater is shallow, operator should spread wastes over land area as described in 1.1.3.2.3 above or transport offsite for disposal.

Grade closure area to the natural upland elevation and treat with 50 lbs./acre of balanced (10-10-10 or 12-12-12) fertilizer to complete closure.

1.1.4 Salts Management.

Soluble salts in waste fluids will detrimentally impact elevated wetland or upland areas if GLC values are exceeded. Excessive salts in waste fluids can induce salinity hazards associated with high osmotic potentials, nutritive imbalances and toxicities to certain plants. Salts can be reduced in elevated wetland or upland areas by mixing with surface soils, leaching of waste/soil mixtures with natural rainfall and utilizing ion exchange mechanisms. Submerged wetland areas will not be impacted as waste fluid salts will be diluted quickly with marsh water immediately after pit closure.

As discussed in 1.1.1 above, analyze pit liquids to determine if liquids can be discharged into waters of the United States. If liquids cannot be discharged, inject liquids underground or transport off site for disposal in accordance with state or federal regulations.

1.1.4.1 Submerged Marsh Areas.

Salt contaminated waste fluids require minimum management only in submerged areas. Close waste fluid pit as described in Section 1.1.1 above.

1.1.4.2 Elevated Wetland Areas.

Salts management in elevated wetland areas is required if waste fluid constituents exceed GLC values. Two closure options assure protection of elevated wetland environments.

1.1.4.2.1

Salts concentrations in waste fluids exceed GLC and sufficient land mass is not available to dilute salt constituents to GLC values. Remove required volume of waste fluid from pit for injection underground or offsite disposal to assure final waste/soil mixture will comply with GLC values. (Waste fluids must be disposed of in accordance with state or federal regulations.) Required volume for removal can be determined by (1) calculating waste/soil ratio necessary to dilute salts to GLC (maximum waste spreading thickness is four inches), (2) calculating volume of waste fluids for spreading over available land area and (3) subtracting calculated volume for spreading from the total pit volume (see 1.1.3.2.3 above for example calculation).

Remove residual waste fluids from pit and spread on available land area to dilute salts to GLC values. Thickness of spread waste fluids must not exceed four inches and should be allowed to dry before surface soil mixing is initiated. Disk dried waste fluids into surface soil to a depth of six inches. Grade treated area to natural elevation of wet- land area and treat with 50 lbs./acre of a balanced (10-10-10 or 12-12-12) fertilizer to complete closure.

1.1.4.2.2

Salts concentrations in waste fluids exceed GLC values, and sufficient land mass is available to dilute salts constituents to GLC values utilizing surface soil mixing. Remove waste fluids from pit. Spread, treat and disk waste fluids over land area as described in Section 1.1.4.2.1 above.

1.1.4.3 Upland Areas

Salts management utilizing land treatment at upland areas will be difficult because dilution of waste fluids to GLC values may not be feasible. Crops and vegetation at upland areas are also highly sensitive to salts and most landowners are reluctant to allow spreading of saline materials on their property. Land treatment of saline waste fluids must be planned and carried out carefully or damage to valuable land can result. If waste fluid salt constituents exceed GLC, land treatment should not be conducted on farmland or residential land, unless the operator is prepared to conduct intensive management. Otherwise, saline waste fluid pits located in these sensitive areas should be buried on site or removed for off site disposal.

Prior to spreading or burying waste fluids, operators should inject as much of waste fluids as possible underground, unless permission from state agencies cannot be obtained. Two closure options assure protection of the upland environment.

1.1.4.3.1

Salt constituents in waste fluids exceed GLC and sufficient land mass is not available for reducing salts concentrations to GLC values. Saline waste fluids should be mixed with burial trenching materials to reduce salts concentrations to GLC. If salts concentrations cannot be reduced to GLC in the waste/soil mixture prior to burial, low permeability soil should be placed over trench areas to prevent leaching of salts to groundwater. The top of the low permeability soil layer should not be located within five feet of the ground surface after closure. Saline waste fluids should be removed for off site disposal if groundwater is shallow. Grade burial area to natural upland elevation and treat with 50 lbs./acre of a balanced (10-10-10 or 12-12-12) fertilizer to complete closure.

1.1.4.3.2

Salts constituents in waste fluids exceed GLC, and sufficient land mass is available for reducing salts concentrations values. This section does not apply to farmland or residential land where saline waste fluids should be buried as in Section 1.1.4.3.1 above or removed for off site disposal.)

Land necessary for reducing the salts concentrations to GLC can be determined easily. First, calculate waste/soil dilution ratio necessary to reduce waste fluid constituents to GLC. For example, an operator wants to land treat a waste fluid with an ESP of 30%. Two inches of waste fluid mixed with four inches of native soil will reduce ESP to 10% which is below GLC. Secondly, calculate land area necessary for diluting waste fluids to GLC. Knowing that only two inches of waste fluid can be spread onto the land and knowing the volume of waste fluids in the pit, the land area can be easily calculated.

Spread waste fluid over calculated land area and grade to the natural land elevation. Once waste fluids have dried, spread calcium sulfate over the treatment area and disk to a depth of six inches, Section 1.4, of this Appendix . Follow treatment with 50 lbs./acre of balanced (10-10-10 or 12-12-12) fertilizer to complete closure.

1.2 LIME AND SULFUR CALCULATIONS

Lime is used to neutralize acidity and control metals. Sulfur is used to control excess alkalinity (> pH 9 in uplands or wetlands).

The amount of lime or acid required to raise or lower the pH, respectively, is determined empirically. One titrates acid samples with 0.04 N $Ca(OH)_2$ and basic samples with 0.1 N HCl to a neutral pH.

In actual practice one uses CaO as a liming agent in wetlands and agricultural lime $CaCO_3$ in upland pit position.

$$\text{Lime Requirement (LR)} = \frac{(\text{mis.titrant})(0.04\ \text{N}\ Ca(OH)_2)}{\text{WT. OF SAMPLE}} \times 100$$

$$\text{LR} = \text{milliequivalent (meq)/100g}$$

where:

$$1\ \text{meq/00g} = 1{,}000\ \text{lb. CaC0}_3\text{/acre - 6 inches}$$
$$1\ \text{meq/100g} = 560\ \text{lb. CaO/acre - 6 inches}$$

When liming with $CaC0_3$ one generally doubles the LR rate to take into account the lower efficiency of Ag-lime in the field.

$$\text{Sulfur Requirement (SR)} = \frac{\text{(mis. titrant(0.1 N HCl))}}{\text{wt. of sample}} \times 100$$

$$\text{SR} = \text{meq/100g}$$

where:

$$1\ \text{meq/100g} = 320\ \text{lb. sulfur/acre - 6 inches}$$

In submerged wetlands, pit liquids less than pH 6 are treated with CaO at the rate of 400 lbs./acre-inch or 27,000 gal. The lime serves as a flocculating agent for suspended colloids and precipitating agent for solubilized metals.

1.3 NITROGEN CALCULATIONS

Nitrogen is added to mud pit materials when needed to adjust C:N values. A ratio of 150:1 C:N is desirable. One does not exceed 400 lbs. actual N/acre - 6 inch in a single application regardless of the nitrogen required. Ammonium sulfate is usually the fertilizer recommended. It is more efficient, and less likely to cause pollution problems, particularly in wetland pit environments. Carbon is based on oil and grease values, with the assumption that oil and grease is 78% carbon.

Example:

A mud pit contains 5% oil and grease.

$$\text{5\% oil and grease} = \text{3.8\% Carbon}$$
$$= \text{38,000 ppm Carbon}$$

C:N of 150:1 is desired

$$\frac{\text{38,000 ppmC}}{150} = \frac{\text{x ppm N}}{1}$$

$$\text{ppm N} = 253$$
$$\text{ppm N} \times 2 = \text{lb. N/acre}$$

One calculates a nitrogen requirement of 506 lbs./acre for a waste fluid pit containing 5% oil and grease. The next step involves an adjustment for actual oil and grease once distributed in the field by mixing with existing surface soil. One generally assumes a dilution factor of 2 in submerged wetlands and 5 in elevated wetlands or uplands. Thus for a submerged wetland siting 5% oil and grease requires approximately 253 lbs. N/acre.

253 lbs. N/acre = 1,204 lbs. [$(NH_4)_2SO_4$]/acre

An elevated wetland or upland siting for a pit containing 5% oil and grease requires approximately 100 lbs. N/acre.

100 lbs. N/acre = 476 lbs. [$(NH_4)_2SO_4$]/acre

= 303 lbs. NH_4NO_3/acre

1.4 CALCIUM SULFATE CALCULATIONS

The gypsum requirement is related to the sodium saturation (ESP) in that it represents the calcium needed to exchange and reduce sodium to below 15% of the cation exchange capacity (CEC). Gypsum or sulfur are added to sodic soils to improve or ameliorate the poor physical characteristics (lack of soil structure and water impermeability) associated with ESP values in excess of 12-15%.

Each milliequivalent (meq) of exchangeable sodium in excess of 15% of the CEC requires approximately 1.7 tons of gypsum or 0.3 tons of sulfur per acre-foot of soil. Soluble calcium and magnesium reported in meq/100 g soil are subtracted from the calculated gypsum requirement.

Example:

A soil has a CEC of 35 meq/100 g and an ESP of 20%. It is desired to have an ESP less than 15%. Twelve is commonly used for calculation purposes. The desired ESP change is 8% (20-12%).

ESP measured - ESP desired = Δ ESP required Gypsum requirement, meq/100 g Δ ESP x CEC

= 8% x 35 = 2.8 meq/100 g

Gypsum requirement (tons/acre-foot) = Gypsum requirement, meq/100 g × 1.7 = 2.8 meq/100 g × 1.7 = 4.76 tons

A pit containing 1.0 acre foot of mud then requires 4.76 tons of gypsum applied in broadcast fashion to spread over treatment area after drying and prior to mixing with surface soil.

1.5 Pit Closure Design Examples

Two examples are provided in this section to show persons how to use Figures 8-1, 8-2 and 8-3 for designing pit closures.

Example 1

Given: Pit size - 100′ x 100′

Mud depth 2′

Land Classification elevated wetland.

Land available for spreading wastes l acre. Determine land area necessary for spreading.

A. Refer to **Figure 8-1**

B. Collect pit liquids and analyze. Discharge, inject or transport liquids offsite as dictated by analysis and regulations.

C. Sample and analyze pit waste fluids and native soil for GLC.
   Waste fluid analysis shows following constituents exceed GLC:
   Organics - 25%
   EC - 60 millimhos/cm
   Barium - 5,000 ppm
   Native soil analysis shows:
   Organics - <1%
   EC - 7 millimhos/cm
   Barium - 500 ppm

D. Pit Closure Diagram, **Figures 8-1, 8-2 and 8-3**, refers operator to Section 1.1.2, 1.1.2.2, 1.1.2.2.1 for metals, Section 1.1.3, 1.1.3.2, 1.1.3.2.3 for hydrocarbons and Section 1.1.4, 1.1.4.2, 1.1.4.2.1 for salts in this Appendix.

E. Metal concentration does not exceed GLC "Maximum Value before Closure". Minimum management can be used since levee materials can be utilized to reduce metal concentration to GLC.

Organics must be diluted below 10% prior to biodegradation, Two inches of waste fluid mixed with four inches of soil will reduce waste fluid concentration to 9%.

Pit Storage Capacity - 150,000 gallons

Two inches of waste fluid spread on 1 acre = 54,000 gallons.

$\frac{54{,}000}{150{,}000} \times 100 = 36\%$ of waste fluids can be treated on site.

Sixty four percent of waste fluids must be injected underground or transported off site.

EC must be diluted below 50 to comply with GLC. Three inches of waste fluid can be spread over one-acre site to accomplish reduction to GLC, but hydrocarbons must be limited to 2 inches of spreading as discussed above.

Organics is the land limiting constituent in this case. If organics are diluted to 9% as calculated above, barium and salts will also comply with GLC after closure.

F. Remove 65% of waste fluids from site (waste fluids should not be buried in wetland areas); spread remaining waste fluids 2 inches thick over 1 acre area. Allow waste fluids to dry.

G. Apply 900 lbs./acre of ammonium sulfate as calculated in Section 1.3 of this Appendix.

H. Disk treatment area to six inches and grade to natural wetland elevation.

I. Follow initial treatment with 2 ammonium sulfate applications of 450 lbs./acre at 8- and 16-week intervals, Section 1.1.3.2.1.

J. Include a balanced (10-10-10, 12-12-12) fertilize treatment of 50 lbs./acre with last ammonium sulfate treatment to complete closure.

Example 2

Given: Pit size - 100′ x 100′

Mud depth 2′

Land Classification Uplands Farmland
Land Available for Spreading 2 acres
Groundwater Depth 100′

A. Enter **Figure 8-1 through 8-3**

B. Collect pit liquids and analyze, discharge or transport liquids off site as dictated by analysis and regulations.

C. Sample and analyze pit waste fluids and native soil for GLC.
   Waste fluid analysis shows following constituents exceed GLC:
   ESP 30%
   EC 30 millimhos/cm
   Barium 6,000 ppm
   Native soil analysis shows:
   ESP 2%
   EC - 2 millimhos/cm
   Barium - 200 ppm

D. Pit Closure Diagram, Figures 8-1 through 8-3, refers operator to Sections 1.1.2, 1.1.2.3, 1.1.2.3.1, for metals and Sections 1.1.4, 1.1.4.3, 1.1.4.3.1 for salts in this Appendix. Because salts exceed GLC, it is recommended waste fluids be buried on site in a farmland area. Metal concentrations must be diluted below 2,000 ppm to comply with GLC. Waste fluids must be mixed 1 part waste fluid with 2 parts soil to accomplish this reduction prior to burial.

Salts concentrations must be diluted below 15% ESP to comply with GLC. Waste fluids must be mixed 1 part waste fluid with 2 parts soil to accomplish this reduction prior to burial.

E. Dig trenches 10-15 feet in depth. Mix trenching soil with waste fluid, 1 part waste fluid with 2 parts soil and bury. Assure top of waste/soil mix in burial trench is below ground surface at least 5 feet.

F. Grade burial area to natural upland elevation and treat with 50 lbs./acre of a balance (10-10-10 or 12-12-12) fertilizer to complete closure.

APPENDIX V

FIELD SAMPLING FOR LOCATING GROUNDWATER TABLES

INTRODUCTION

Depth to groundwater is a concern for NOW management techniques utilizing burial and vertical dilution. The concern emanates from the potential for materials buried or distributed in a subsurface profile to be brought back to the surface or further concentrated by a rising water table. For this reason it is suggested that the depth of the trench or burial vault be maintained a minimum 5 ft above the seasonal high groundwater table. The state of Louisiana restricts this activity by statute (Amendment to Statewide Order No. 29-B).

Perhaps more obvious is the potential for direct dissolution of waste constituents into groundwater on contact. The concern is possible contamination of shallow aquifers used for human and animal consumption.

Weathered soils in humid regions may have a saturated zone develop at a discontinuous lithic contact during periods of high rainfall. The discontinuity could be either textural (i.e. sand over clay) or structural (i.e. weakly aggregated over massive) in nature. The saturated zone developed is called a perched water table. A perched water table is separated from permanent groundwater by an unsaturated or "vadose" layer. A sufficiently thick unsaturated zone presents no limitation for NOW management via burial techniques.

This appendix advances a procedure for determining moisture regimes with depth. The procedure utilizes incremental physical and moisture analyses on continuous cores to describe a given site in vertical profile. Parameters analyzed include 15 bar moisture, 1/3 bar moisture, the existing moisture content, and porosity.

The 15 bar moisture content describes a condition in which water is bound to soil particles so tight that it is unavailable to plants. Hydraulic gradients and water transport mechanisms are controlled by diffusivity and other unsaturated flow phenomenon. It is measured by first saturating the soil with water, then allowing it to come to equilibrium with an applied pressure of 15 bars.

The 1/3 bar moisture content represents the maximum volume of water a soil can hold before free liquid subject to gravity flow forms. Soil is saturated, then allowed to come into equilibrium with an applied pressure of 1/3 bars.

Moisture as received is a gravimetric determination of water content existing at the time the sample was collected. Porosity is simply the the percentage of open spaces or voids in a soil and represents the maximum moisture level that can be attained.

Characteristic profiles developed by plotting the various parameters as a function of depth describe the site hydrologically. Characteristic profiles vary from year to year and between seasons within years. For this reason it is best to collect core samples at a season corresponding to minimum evapotranspiration and maximum rainfall (determined from published climatological data). In the Southeast and Gulf Coast States the seasonal high groundwater table usually occurs in late December or early January.

This appendix was written to provide a technique whereby one may define the nature of subsurface moisture characteristics, particularly, the differentiation between a continuous groundwater table or aquifer and a perched or seansonal water table. It also defines the depth interval of an unsaturated zone with potential for receiving NOW materials. Obviously too thin a zone limits the utility of burial making other techniques more practical from an environmental and economic viewpoint.

The technique of describing subsurface hydrological characteristics from soil core data is advanced as an alternative to expensive geohydrological investigations utilizing groundwater monitoring wells. These investigations do not adequately describe perched water table conditions or soil vadose.

PROCEDURE

Soil Corings

Continuous soil cores are collected at the site to receive NOW materials. Several cores are taken to describe spatial variability. Cores may be taken seasonally to establish temporal variability. This would be recommended where climatological data is scant or non-existent.

Moisture Equivalents

The 1/3 and 15 bar moisture equivalents are determined on core samples segmented into 1 ft increments. Core samples are then split in half. The existing water content is determined by gravimetric weight loss which is the equivalent weight lost upon drying to constant weight at 105 C. The other half of the core sample is dried, ground and sieved to pass a 2 mm screen. Samples are saturated in contact with a ceramic plate and placed in a cell pressured at 1/3 bar and another at 15 bars. Soils are allowed to equilibrate at the pressures indicated followed by gravimetric moisture analysis of the sample by drying to constant weight at 105 C.

Porosity

Porosity is calculated as percent pore space by means of the following equation:

$$\text{Porosity, \%} = [100 - (\text{bulk density} / \text{particle density}) \times 100]$$

Bulk Density

The bulk density is measured as the ratio of mass to bulk volume. A paraffin clod technique gives an accurate assessment of strongly structured soils having aggregates of suitable size. Clods are often broken out from the core half used to determine existing moisture content. Loose or poorly structured soils are ground and sieved, then placed in a graduated cylinder to about 1/2 full. The soil is settled by gently tapping against the table top and volume occupied recorded. Contents are then weighed with bulk density reported as the mass per unit volume.

Particle Density

Particle density is a measure of the collective density of the solid particles. Dried and sieved materials are weighed into a volumetric flask. The flask is brought to approximately 1/2 its volume with water, followed by gentle boiling and agitation to remove entrapped air. Contents are cooled and brought to volume. The temperature of the water is recorded and flask weighed. Contents are removed from the flask and brought to volume with water and weighed again. Particle density is calculated as follows:

$$Ds = \frac{Dw\,(Ws - Wa)}{(Ws - Wa) - (Wsw - Ww)}$$

Where
Dw = Density of water at observed temperature
Ws = Weight of flask plus soil
Wa = Weight of flask filled with air
Wsw = Weight of flask filled with soil and water
Ww = Weight of flask filled with water at observed temperature

Results are generally presented by plotting the 4 parameters on a single graph, depth on one axis, moisture and porosity on the other. The area on the graph between the 15 bar moisture and the porosity is the range of moisture levels that could occur.

INTERPRETATION

At the point where percent moisture (as received moisture of the core sample) exceeds the 1/3 bar moisture equivalent a condition of free water exists. If the percent moisture exceeds 1/3 bar and remains in excess, a continuous ground water table is demonstrated **(Figure A-V-1)**.

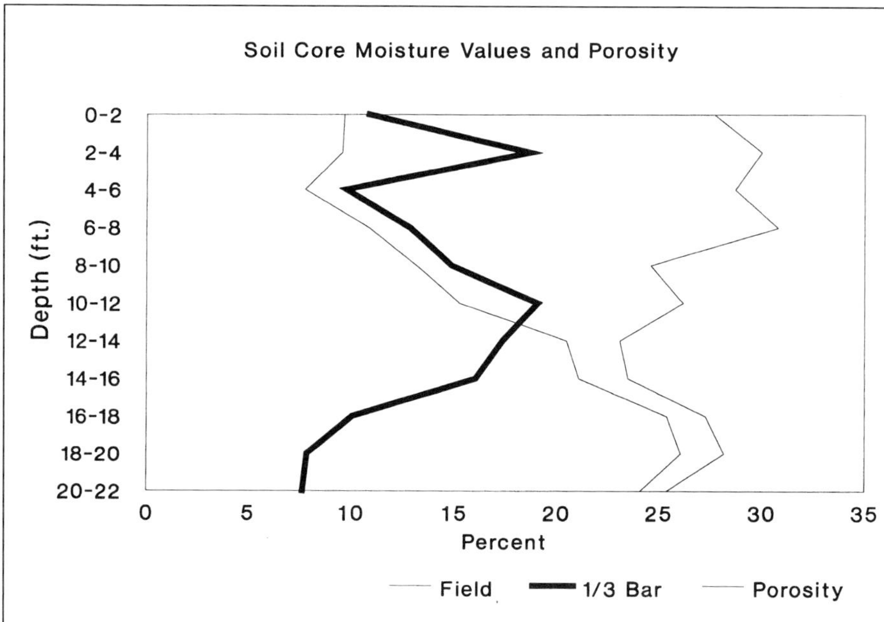

Figure A-V-1 Moisture Profile Demonstrating a Ground Water Table.

In the case where the percent moisture exceeds 1/3 bar at a given depth but recedes to a value at less than the 1/3 bar equivalent deeper in the profile, then a perched water table is demonstrated (Figure A-V-2). Depending upon the site and climatological conditions one may identify several discontinuous zones of saturation. It should be noted that the same physical constraints holding water in the profile serve to retard NOW constituent migration.

Burying NOW materials in a saturated zone in continuous contact with groundwater is a very undesirable condition and should be avoided unless constituents levels meet residential/farming upland soil criteria or the NOW materials are solidified to inert status.

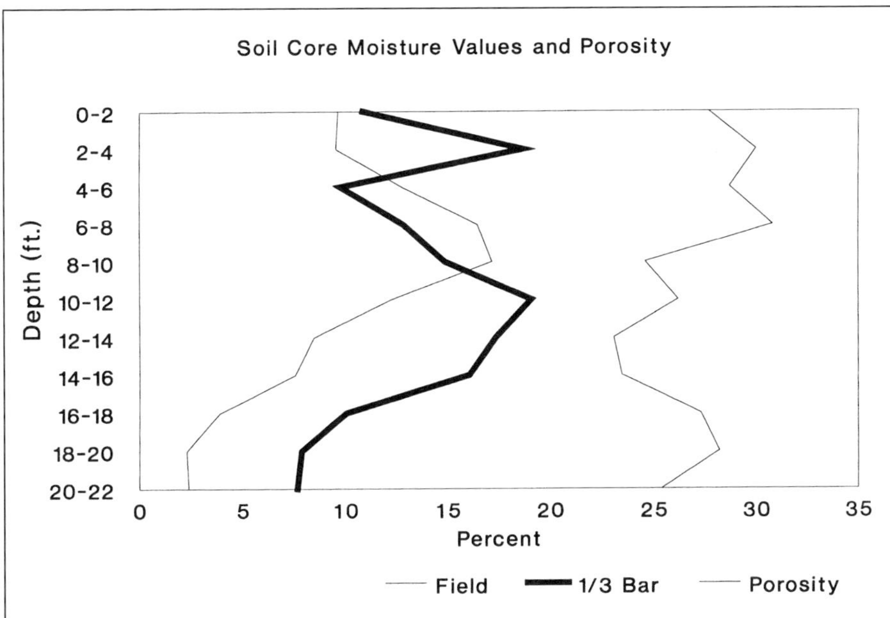

Figure A-V-2 Moisture Profile Demonstrating a Perched Water Table.

APPENDIX VI

RADIATION SAFETY

INTRODUCTION

NORM produces low level radiation, which does not penetrate the skin. However, NORM radiation when ingested or inhaled can cause extensive harm and should be avoided. Implementing the procedures contained in the following Annexes provides reasonable assurance that workers at NORM-contaminated sites are protected while performing a survey.

ANNEX A

TRAINING COURSE FOR INDIVIDUALS PERFORMING NORM SURVEYS

The folowing outline describes the subjects individuals should demonstrate competince in prior to performing a NORM survey.

1. Fundamentals of Radiation Safety
   - A. Characteristics of radiation
   - B. Units of radiation dose and quantitiy of radiation
   - C. Levels of radiation from sources of radiation
   - D. Methods of minimizing radiation dose

     1. working time
     2. working distance
     3. shielding
     4. respiratory precautions
     5. use of anti-contamination clothing

2. Radiation Detection of Instrumentation to be Used
   A. Use of radiation survey instruments
      1. operation
      2. calibration
      3. limitations

   B. Survey techniques

   C. Use of personnel-monitoring equipment

3. The Requirements of Pertinent State Regulation

ANNEX B
WORKER PROTECTION PLAN

Expanding the following outline provides a basic NORM worker protection plan for oilfield operations.

1. The posting of signs pursuant to inform personnel when they are entering a NORM-contaminated area.
2. Procedures to prevent eating, drinking, smoking, and chewing in areas where work is being performed on NORM-contaminated equipment or where contaminated soil is being handled.
3. Procedures to avoid skin contact with NORM-contaminated scale, solids, and sludges by the use of protective clothing such as gloves, coveralls, rubber boots, and eye protection.
4. Procedures to ensure personnel will thoroughly wash body parts that may be been in direct contact with NORM-contaminated materials before eating, drinking, smoking, or leaving the work area.
5. Procedures for whole body monitoring of personnel in restricted area including a description of the survey instrumentation used.
6. Procedures to minimize the number of personnel in the contaminated area
7. Other operational procedures (i.e., openings on contaminated equipment or containers should be sealed when stored or handled, tubulars capped on both end, etc.).

For operations having the potential to produce NORM contaminated dusts (i.e., cutting, grinding, sand-blasting, welding, drilling, polishing, or handling soil) or when loose contamination is suspected, the following additional precautions should be taken:

1. The use of a respirator appropriate for radioactive particulates should be worn.
2. Safety glasses should be worn for eye protection.
3. Activities must be conducted in well-ventilated areas to which access has been restricted.
4. Ground covers should be utilized to the extent possible to contain contaminants and facilitate clean-up.
5. The need for personnel monitoring and bioassay shall be evaluated and provided if necessary.

In addition to the general guidance given above, there may be industrial operations such as vessel entry, dismantling of equipment, refurbishing of equipment, or transportation, which may require additional precautionary procedures which should be included in the worker protection procedures.

INDEX

A

B

C

D

M

N

O

P

Q

R

S

T

U

V

W

Y

Z